Maximilian Schele de Vere

Wonders of the Deep. A companion to Stray leaves from the Book

of Nature

Maximilian Schele de Vere

Wonders of the Deep. A companion to Stray leaves from the Book of Nature

ISBN/EAN: 9783337025113

Printed in Europe, USA, Canada, Australia, Japan

Cover: Foto ©berggeist007 / pixelio.de

More available books at **www.hansebooks.com**

WONDERS OF THE DEEP.

A COMPANION TO

STRAY LEAVES FROM THE BOOK OF NATURE.

BY

M. SCHELE DE VERE.

NEW YORK:

G. P. PUTNAM & SON, 661 BROADWAY.

1869.

CONTENTS.

WONDERS OF THE DEEP.

I.

FABLES AND FACTS.

"Whence and what art thou, execrable shape."—Milton.

THE great sea has its mountains and its deep valleys, with forests of weird, waving plants on the
former, and, far down in the dark dells, masses of dismal
débris, wrecks of vessels, and decaying bodies of men.
There lies, half-covered with a crust of lime and hideous
green slime, an ancient gun shining in sickly green ; here,
half-hidden, a quaint box filled with gold that was picked
up amid the snows of the Peruvian Andes, and scattered
over all a motley crowd of oddly-shaped shells. The
empty skull of an old sea-captain has sunk down close
to the broken armor of a huge turtle, and a deadly harpoon rusts and rots by the side of the enormous tooth of
a walrus. Still farther down, " in the lower deep of the

lowest," lie countless bales of Indian silks, in which
large schools of fish dwell peacefully now ; and over all,
the silent currents of the ocean move incessantly to and
fro, while millions of jelly fish throng every wave to feed
the giant whales, and immense hosts of herring dash
frightened through the waters to escape the voracious
shark. Not only mountains and valleys, however, break
the apparent monotony of the vast deep, but all that the
surface of the earth can present of picturesque beauty
or horrible hideousness is repeated below. In one place
the waters foam and the waves break without rest or repose against oddly-shaped cliffs, which do not rise sufficiently high to be seen above ; in another, they wash
slowly and sadly against a wide desert of white sand.
Where lofty mountains rise from the depth to a height
not inferior to that of the tallest of Alpine summits, and
vast forests of sea-tang clothe them in brilliant green, the
sea circles mournfully all around in ceaseless windings,
while farther on, where the valley sinks into the very
bowels of the earth, and eternal darkness covers all with
its mysterious mantle, the waters themselves are hushed
and apparently motionless, as if awe-struck by the unbroken silence and the unfathomable night below.

The earth, it has been said, is one vast graveyard, and
man can nowhere put down his foot without stepping on
the remains of a brother. This is not less true with regard to the ocean. It is an ever-hungry grave, in which
millions and millions of once living beings lie buried,

and new hosts are added from year to year. It is the stage on which murder and maddest conflict are going on without ceasing. Immeasurable hatred dwells in those cold, unfeeling waters—and yet for a good purpose, since it is only through this unceasing destruction and change that life can be maintained in the crowded world that dwells in the "waters below the earth." The sea has its lions, its tigers, and wolves, as well as the earth above, its crocodiles and gigantic snakes, which daily sally forth to seek their prey and murder whole races; it has its medusæ and polypi, which spread their nets unceasingly for smaller fry; while whales, and their kindred, swallow millions of minute beings at a single gulp, swordfish and sea-bears hunt the giants of the lower world, and wretched parasites wait their opportunity to enter the fatty coats of huge monsters. Every thing is hunting, chasing, and murdering, but there is heard no merry "Tally Ho!" no war-cry encourages the weary combatants; no groan of pain, no shout of victory ever breaks the dread silence. The battles are fought in dumb passion, and no sound accompanies the fierce conflict but the splash of foaming waters and the last spasmodic effort of the wounded victim.

Can we wonder, then, that from time immemorial the sea has been peopled by the learned and the ignorant alike with marvels of every kind? It is the good fortune of travellers, and especially of those who "go down to the great deep," to be either determined that all they have

1*

seen must needs be unique, unheard of, and marvellous, or disposed to follow the *nil admirari* doctrine, and to insist that they have never met with any thing which was not perfectly familiar to them already from previous knowledge, or at least very easily accounted for by their superior mind. Vanity induces the former to magnify, self-love teaches the other to diminish all they have seen, and thus littleness of mind unfits both for correct observation and candid reports. The ancients, with their very limited knowledge of the sea and its life, very naturally transferred the features of the world above to that below the waters, and their lively imagination peopled the ocean with all the animals that were familiar to their eyes. There were sea-horses and sea-lions, poisonous sea-hares and ravenous sea-wolves, sea-swine, and even sea-locusts. There was the Chilon, with a man's head, living frugally on nothing but his own viscous humors; and there was the Balena, not so very like to a whale, and most cruel to its mate. There were those real wonders of the sea, the Dolphins, who swam about with their babies at the breast, and their eyes in their blade-bones, who dug graves for their deceased parents and friends, followed them in funeral procession, and buried them in submarine graveyards out of the way of the fishes. There was that strange fish, the Dies, with two wings and two legs, which in the perfect state lived only for a day. There was the Phoca, another oceanic brute, who was perpetually fighting with his wife until he killed her.

Always remaining in the same spot, where he had murdered one wife, he disposed of her body and took another,
thus playing Henry VIII. to a series of wives, until he
either died himself or found a mate who was a match for
him.

But these were only the teachings of that despised
science, Natural History. Of far greater interest to the
nation, and of deeper concern for their future welfare,
were the legends of the god-like shepherds, a Proteus, a
Nereus, and a Glaucus, who tended the aquatic flocks of
Neptune, and were endowed with marvellous powers.
We have all seen those classic bas-reliefs, in which the
great Poseidon is accompanied by monsters, half-horses,
half-fishes, while others wear partly the semblance of
men, and blow trumpets made of huge shells with such
terrible force, and such fearful sound, that their notes
calmed the stormy sea! These hippocampi, sometimes
provided with cloven feet and long tails, and then again
covered all over with scales, and of the color of the sea,
were favorite forms with some of the greatest sculptors
of antiquity, like Myron and Scopas; and yet it is held
doubtful whether they were altogether the offspring of
poetic or artistic inspiration. Naturalists, as well as artists, have been led to think that extraordinary creatures
of somewhat similar shape may have really peopled the
seas in ancient times, and that they, like the giants of
old, which are now reduced to the moderate proportions
of our day, may have dwindled down into the diminutive

hippocampi which abound in Southern waters.. These little sea-horses, as they are familiarly called, have the perfect form of a horse's head, with prominent round eyes, and a steep, straight brow, while the gills float in exact imitation of a mane from the proudly-arched neck. They could not have copied the walrus, as that strange, monstrous animal, of which we shall presently have to say more, lives only in polar regions, to which the ancients did not have access. When the Tritons, on the other hand, were represented in human shape, they belonged, of course, simply to the realm of fables. And yet strong and frequent evidence is given by ancient authors of the real existence of beings whom they resembled. Demostratus, for instance, relates that such a Triton was still to be seen, imperfectly embalmed, in a temple of Bacchus at Tanagria.

It seems that there had been enmity for generations between these strange children of the sea, and the good people of Tanagria. One of the Tritons had been in the habit of coming forth every night from the waters to steal the cattle on shore, and all efforts to catch him on the part of the dwellers there had long been in vain. At last they placed a vessel, filled with strong wine, on the brow of a steep hill. When the Triton came, according to his custom, he noticed the vase, and was curious to ascertain its contents. He tasted, he liked it, and drank till he fell fast asleep on the edge of the precipice. During his disturbed slumbers, he rolled over and fell from the great height upon the rocks below, where the Tanagrians lay in wait, and wreaked their vengeance on the formidable robber.

Pausanias saw a smaller Triton at Rome, and from that time the annals of all countries of the world abound with strange legends of uncouth, horrible beings, born and bred in the sea, who entered into ill-fated relations with men, and almost invariably contrived their ruin. The White Lady of Scotland, the Nix or Undine of beautiful German lore, the Merminne of the Netherlands, and the Nech—our Old Nick—of the dismal North, are all children of the marine monsters of antiquity. Among the latter, some were great favorites with poet and priest, and their memory survives to our day. Thus the Ocean itself was represented as the son of the Heavens and the Earth, and the first of that gigantic race of Titans who stormed the abode of the gods, but the only one who did not join in the revolt of Saturn. How the briny deep was made to differ from the vast lakes with sweet water, their religion did not tell; but the distinction was made at an early date, for Hesiod already tells us that "nine-tenths of the waters of the ocean, passing under the earth across dark night, fall in silvery showers upon the bed of the waves, around the earth, and on the back of the seas. One-tenth only, to the great injury of the gods, escaping from a lofty rock, forms the waters of the Styx, and by it the Immortals are fond of swearing."

Among the vast offspring of the ocean, again, the

Nereids stand foremost by their number and by their beauty. They were all fair young maidens, nearly naked, and are often seen in the frescoes of Pompeii, and elsewhere, in most graceful positions, reclining on the back of sea-horses, or giving drink to thirsty monsters of the deep. It was only when the taste of artists became corrupt, and the fancy of men ran riot amid Eastern fictions, that they were represented as ending in fishtails, and as having hair of the color of the sea. Another sea-god, marrying the Muse of Lyric Poetry, was presented by her with three daughters, the Sirens, whom he called Blanche, Harmony, and Virgin Eye; but, unfortunately, he lost them soon after, when the infuriated Ceres punished them for having allowed the carrying off of her daughter Proserpina, and changed them into monsters, half women, half birds. The unfortunate maidens fled in despair, and hid themselves in the Islands which dot the waters between Sicily and Italy. But even there the curse pursued them still, for the decree had gone forth, that they were to die if ever man should pass them without stopping. Behold, now, the poor metamorphosed beauties straining their sweet voices, and blending them with the softest notes of their instruments, in order to attract hapless seamen, and to draw them into ruin. Surely, the ancients felt that sea and land alike are welcome stages for the fatal skill of the coquette! Only once the sad sirens were foiled in their attempts to win and to ruin the children of men. It was when the *élite* of Grecian heroes sallied forth on their great expedition in search of the Golden Fleece—in reality, a company of daring adventurers, who went to take possession of the gold mines in the Ural Mountains—and passed close to the islands on which the wretched sisters were living. They came down to the steep sides of the precipices, they displayed their unequalled charms, and sang their sweetest to cast their spell over all their senses. But Orpheus, who had joined the merry company with his lyre, raised his own sweet voice, and soon they were forced by its wondrous power to listen in their turn, and to let the Argo pass unharmed. Perhaps the godlike nature of the great singer was pleaded in their behalf, for they survived the future; and it was only when cunning Ulysses used the coarse trick of filling the ears of his companions with wax, and thus rendered them insensible to their enchantments, that they paid the penalty, and were changed into rocks. Even then one of them survived; for the compassionate waters refused to bury her; they sent her back to the surface, and she became fair Naples, the city of magic beauty, where so many have died from over-enjoyment, obeying literally the ancient saying: *Vedi Napoli e poi muori!*

Pliny seems still to have been in doubt as to the real existence of these marine monsters; at least, he defends himself against the suspicion of believing in them with an earnestness which goes far to prove the lingering doubt. "I do not believe in sirens," he says in his book

on birds, " although Dino, father of Clearchus, a famous
author, affirms that they exist in India and tempt men by
their song, in order to tear them in pieces when they are
asleep." In another place, again, he believes them to
have been real fish, which recalled, in a vague manner, the
features of human beings, and states that several such had
been taken on the coast of Gaul.

The assertion is, strangely enough, supported by later
evidence; for other portions of the earth, and later ages,
have all faithfully repeated the legend, and pointed to
actual beings in the sea as proof of their truth. Have
not even the Arabs—who either ignore the sea altogether,
or hate it as cursed by their great prophet—their weird
beings, half men and half ostriches, who live on desolate
islands, and devour the bodies of shipwrecked mariners
brought to their rocks by the friendly waves? Near
Rosetta and Alexandria, in Egypt, the waters are peopled
with still stranger creatures, poetically called the Fathers
of the Fair, who come only on shore for peaceful purposes,
walk quietly about to enjoy the sweet air of heaven and
the perfumes of flowers, and then return reluctantly to
their dark homes in the great deep. A hundred of them
were once captured, but they uttered such very sad sighs
and unbearable groanings, that the hunters released them,
and saw them plunge with delight into the cool waters.
The Old Man of the Sea is familiar to all our readers
through the Arabian Nights; but it is less generally
known that he occasionally appeared near Damascus, and

then promised a good harvest to the Syrians; the people were so grateful to him for his benevolence, that they caught him once and married him, fishtail and all, to a fair daughter of the land. The monster was well content, but not so the farmers, for his happy influence had left him as soon as he had found his master in his wife. Other Arabic authors tell us even the religion of one of those marine beings; he is called by them the Old Jew, and appears on the night preceding the Sabbath, with his white hair and shaggy coat, on the surface of the Mediterranean, and remains there, swimming about, plunging, and jumping high, and following the vessels as they pass near his home, till the Sabbath is over, and he sinks once more down under the waters.

These strange beings, reported to have been found or heard of with at least as much accuracy and as frequently as the Sea Serpent of our days, were evidently the ancestors of the mermen and mermaids, the ill-starred, God-forsaken dwellers in the kingdom of waters, the

> " Merman bold,
> Sitting alone,
> Sitting alone
> Under the sea,
> With a crown of gold,
> On a throne;
> And the mermaid fair,
> Singing alone,
> Combing her hair,
> Under the sea,
> In a golden curl
> With a comb of pearl,
> On a throne."

For a time Christian authors loved to revive the fables

of pagan antiquity, or unconsciously repeated the weird
fancies of older nations. Soon, however, certain features
appear in their accounts, which show that they were ·
either reports of real discoveries of marine monsters,
dressed up, perhaps, in somewhat fanciful colors, or at
least new inventions in harmony with the spirit of the
age. The mermen soon cease to be mere monstrosities;
they appear in a form resembling human beings, often
scarcely to be distinguished from the people near whom
they live, whence follows more frequent intercourse and
a closer intimacy between the two races.   One merman,
found on the outermost point of Mauritania and brought
to Spain, is reported as still having been in part a fish;
but Theodore of Gaza already describes the mermaid, of
which he saw several cast ashore on the coast of Greece,
as fair and graceful; one of them he assisted in reaching
the water, and immediately she plunged into the waves
and was seen no more.   Other authors, of such high re-
pute that even the great Scaliger may be mentioned
among them, tell of such wondrous beings, which they
saw themselves or heard of through trustworthy friends.
These accounts were, of course, valued only in proportion to the wonder they excited, and added nothing to
our actual knowledge of the dwellers in the waters.
They led, on the contrary, to new errors, and much
amusement might be derived from the precepts given to
unlucky sailors who should fall in with such sirens.
They were advised to cast bottles into the sea, with

which the monsters would play long enough to give
them time to escape; to stop their ears carefully with
wax and oakum, and to invoke aid from on high against
their enchantments. The great Cabot, so intimately
connected with the history of our continent, furnished
the officers of the first vessel that ever attempted the
voyage to the fantastic Cathay of those days, with a set
of curious instructions. He recommended that prayers
should be held twice every day, and all inventions of the
Evil One, like dice, cards, and backgammon, should be
strictly prohibited. By the side of such excellent suggestions are some of more doubtful morality. Thus he enjoins
upon the officers to attract the natives of foreign lands,
to bring them on board ship, and there to make them
drunk with beer and wine till they had revealed all the
secrets of their hearts. The rules contain at the end a
recommendation " to take good care against certain
creatures which, with the heads of men and the tails of
fishes, swim about in the fiords and bays armed with
bow and arrows, and feed upon human flesh."

The dark North, with its misty, murky atmosphere,
which is reflected in the sombre legends of dismal superstitions, has its mermen above all others. They are
mostly seen when fearful tempests threaten destruction,
or sudden storms bring shipwreck to vessel and sailors
alike. It is but here and there that they are painted in
softer colors. In one of the legends, a famous giant of
the seas, called Rosmer, carries off a Danish maiden of

great beauty; she has to live with him in a great mountain, to which he comes every now and then from his
home in the waters. Her brother, who had sallied forth
to find her and to rescue her, lands at the desert rock,
and is at first in great danger of being slain and devoured by the terrible monster. He succeeds, however, in
pacifying the merman, and serves him faithfully for many
years. At last he obtains leave to return home, and receives, as reward for his services, a large box filled with
gold and precious stones. The giant even condescends
to carry the box himself on board the ship, unconscious
that the cunning maiden has first taken out all the treasures, and then concealed herself in the box, from which
she comes forth as soon as the ship has reached the high
seas.

It may readily be imagined that mariners who set out
on long voyages to distant, unknown lands, with their
minds filled with such images and marvellous stories,
were ready to see sirens and other wonders of the deep
to their hearts' content. Christopher Columbus even,
when sailing along the coast of St. Domingo, met with
three sirens, who were dancing on the water. They had,
however, no sweet songs with which to allure him, and
their silence, combined with their lack of beauty, made
him think that they probably " regretted their absence
from Greece." There can be little doubt that his sirens
were Manatees, huge monsters so called because they
carry their young with their flappers, or finlike hands,

and give them suck on the breast—relations of the great
Dugong of India, the only animal yet known that grazes
at the bottom of the ocean. It has the strange power of
suspending itself steadily in the water, and its jaws are
bent in such a curious manner that the mouth is nearly
vertical, by which means it is enabled to feed upon the
sea-weeds down in the deep, very much as a cow does
upon the herbage in the bright sunlight above. The
Manatees serve to frighten the children of African slaves
even now, when they suddenly rise like " spirits from
the vasty deep," their large, gentle eyes looking anxiously around, and their young clasped tenderly to their
bosom—a favorite position of theirs, which has earned
them, with Spanish colonists, the name of Fish-Women.

The poor Brazilian natives, who still cherish the traditions of their forefathers, fondly believe in the existence
of an immense lake in the interior which contains an
enormous treasure, guarded and watched over by a siren
whom they call the Mai das Aguas. They also believe
still in the accounts given by early discoverers of strange
beings met in their waters. Did not even brave John
Smith, the valiant hero and daring navigator, when he
came near our own continent, see a woman swimming
gracefully near the vessel ? Her eyes were large, beautiful, and full of expression, although rather round, the
nose and ears well made, and the hair long and soft, but
of sea-green color. His heart was near giving way to all
these charms, when the strange being suddenly turned

over, and showed to her disconcerted admirer a forked fish-tail!

Among South American Indians, it seems, tails of mermen are a favorite subject, though here and there these marine monsters are dreaded with instinctive abhorrence. Moravian missionaries have sent home strange reports of these superstitions, and yet found themselves unable, in their desire to honor the truth and to avoid misstatements, to deny positively all ground for these traditions. For not only the natives, but the ministers and agents of the pious Brethren themselves, firmly believed that they had met with men and women who lived in the water. They furnished statements, apparently made in full earnest and godly sincerity, that they had actually seen brownish beings with human faces and long hair rise suddenly from the water, and that the urgent intercession of the Indians alone had kept them from killing the supernatural beings. The natives looked upon them with superstitious awe, and insisted upon it that to kill one of them would be simply to bring dire calamities upon their settlements and the whole race. It must be presumed that we meet here with stray members of those aquatic tribes of Indians who live actually more in the water than on land. Martius, and other travellers, down to our day, tell us that the Indians who dwell near the upper branches of the Paraguay, the Maranhao, and other large rivers, remain for hours and hours in the water, and are such expert swimmers that

they defy the most powerful current, and dive like water-
fowl. A small bundle of leaf-stalks taken from the Bu-
riti palm-tree is all they use ordinarily for their support;
at other times they seize an oar, hold it between their
feet, and use it as a rudder to steer with, and thus swim,
holding their weapons in their muscular arms; or they
leap with incredible agility upon a tree floating along on
the swollen stream, sit down on it astride, and thus cross
in a few minutes the most rapid current. No cayman or
aquatic animal is safe from them, and they fight and
defeat the huge capyvara, and the largest serpent, with
great courage. They fear literally nothing except the
Minhoças, a fabulous creature which is said to live in the
rivers and still waters of Equatorial Brazil, and which
naturalists believe to be either a giant eel endowed with
powerful teeth, or perhaps a large variety of the famous
gymnotus with its galvanic battery. These Canoeiros,
as the Water-Indians are called, are true Ishmaelites;
they are at war with all the other tribes, and are therefore hunted down like wild beasts ; they have no home
and no country of their own, and hence they may very
well have given rise to the fabulous reports of mermen
still rife among the credulous Indians of that continent.

In Germany, where folk lore abounds and superstition
still has its strong hold on the minds of the masses,
gruesome stories are told in the long winter nights of
the Nixen, who dwell in the waters near the coast, in
crystal-clear rivers, under the dark shadow of ancient

trees, and in bright, bubbling wells, in half-hidden glens.
They are the sirens of the sunny South, and even here
the ancient curse seems to follow the ill-fated race. For
here, also, they are condemned to expiate some great and
grievous sin committed by their forefathers, and to suffer
long and miserably. As the whole creation groaneth,
however, these sorrowful beings also yearn to be released, and of this longing many a touching tale is told
in German legends. Thus one of them tells us, that
the children of a Protestant minister were once playing
on the banks of a river, when they saw a Nix rise from
the waters, who, thinking himself unobserved, began to
sing and to play on a strange, but ineffably sweet instrument. With the cruelty common to children, they at
once rushed upon him and reproached him for his merri-
ment, adding that as he was nothing but a condemned sin-
ner, he had much better weep over his eternal wretched-
ness. The poor water-sprite, taken by surprise and dis-
tressed beyond measure, broke into tears; and the youthful tyrants, delighted with their success, went home to
tell their father what had happened. But they were
badly received here, and told that they had acted very
wrongly and must return at once and comfort the poor
being whom they had so grievously afflicted. They ran
back, and as soon as they saw the Nix they cried out to
him not to weep any longer, since their father had said
that the Lord had died even for him, and he also might
hope to be forgiven hereafter. Thereupon the poor Nix

dried his tears, recovered his cheerfulness, and played
with them all day long.

Holland, with its wondrous bulwarks and its lifelong
conflict with the sea, abounds naturally in stories of every
kind, in which mermen and mermaids play a prominent
part.　Sometimes they meet the intrepid sailor out on
the high sea and sing of his joyous return, or warn him
of his approaching end ; at other times they come on
shore, make themselves useful in a thousand ways, and
vanish only when they are ill-treated or laughed at.
There is hardly a town on the seacoast which has not its
own legend of this kind; but generally the men are less
interesting than the maidens, since the latter are prophets
and play a prominent part in the sad history of that country.　Such was the mermaid that once frequented the
waters near Zevenbergen, a fortified town with massive
walls and lofty towers, in which dwelt thousands of opu-
lent citizens with their wives and children.　But the people were as wicked as they were rich; and professed to
believe neither in heaven nor hell.　One fine day the
siren appeared in company with a sister mermaid, and
with solemn, tearful voice both began to sing :

"Zevenbergen must perish,
And the tower of Lobbekens alone shall remain."

In spite of this warning the inhabitants continued their
riotous living and sinful profanity.　In a dark November
night of the same year a fearful tempest arose; the wind
blew from the northwest, and with such terrific force that

2

the dykes gave way under the overwhelming pressure of the waters, and the Saint Elisabeth, as the inundation was called, overwhelmed not less than seventy-two towns and villages. Among these was the unfortunate town of Zevenbergen, and so thorough was its destruction in the deep waters, that when the morning broke, and people came from a distance in boats, they saw far beneath them the ruins of houses, and nothing standing but the one lofty tower of Lobbekens. Thus the prophecy of the mermaid had become true. Fortunately, man has triumphed over the evil prophet and the element alike. By an immense outlay of capital and the incessant labor of long years, the whole vast region has been once more laid dry, and from the midst of polders, or dyked meadows of surpassing fertility, there rises now a new town of Zevenbergen, richer and wiser than the doomed village of former days.

Holland is also the land which has originated the very peculiar faith in legends of sea-knights and sea-bishops, some of whom were captured from time to time and exhibited in the large cities. They were found afterwards in all the northern seas, and the works of those ages, down to the latter part of the sixteenth century, contain generally one or two so-called faithful likenesses of these very curious monsters of the deep. In 1305 already a sea-knight was caught out in the open sea to the north of Dockum, and carried from town to town; his fair appearance, and especially the complete suit of armor which he wore, excited universal admiration; but he died, unfortunately, in the third week, at Dockum.

A work of great scientific merit, and published as late as 1534, contains an engraving representing a sea-monk, whom the author, Rondelet, heard of in Norway, where it had been taken after a fearful tempest. It has the face of a man, but rough and repulsive, a bald, smooth head, the cowl of a monk hanging over the shoulders, two long fins instead of arms, and a body ending in a huge double-fluked tail. Other monks of the same kind appear in similar works, sometimes wearing a bishop's habit and mitre, and one of them is reported to have been sent in 1433 from the Baltic, where he was captured, to the king of Poland. The poor creature, however, refused steadfastly to utter a sound or to take any food; the king, moved with compassion, ordered him to be carried back to the sea, and the monster no sooner saw his own element than he gave signs of exuberant joy, leaped into the water and was never seen again. It may be added, that the Protestants made great capital out of these marine dignitaries of the church, and hence gave rise to the suspicion that the whole race of sea-monks and sea-bishops was artistically produced as a quaint revenge which the Reformation took on the persecuting Church of Rome.

The explanation is perhaps only an afterthought, but, as the proverb has it, that there is no smoke without fire, so here also, these countless and persistent traditions

contain their grain of truth, which has been only half
hid in a bushel of falsehoods. The fact is, that these
fables could never have been invented, much less authen-
ticated, even after the imperfect manner of early ages, if
there were not certain animals living in the great deep
which possess sufficient likeness to the human form to deceive careless and superstitious observers. If there are
no real tritons and sirens to be met with in our waters,
such as we see in ancient sculptures, or the coats of arms
of noble families, there are at least seals and walrus, sea-
lions and sea-cows, and similar monsters, whose faces
and gestures, as seen on the surface of the waters, recall
forcibly the features and movements of men. Unscrupu-
lous cheats have occasionally taken great pains to
manufacture actual sirens, and their remains are to this
day carefully preserved in many a museum of European
cities; like the well-known sirens of Leyden and the
Hague. Nor is our own time exempt from these attempts
to profit by the credulity of men. At the beginning of
this century, a crafty fisherman on the coast of India
skilfully joined the body of an ape to the lower part of
a large fish, and dressed up the whole affair so cleverly
that even experienced men were taken in, and bestowed
much time and long research upon the extraordinary
being. As the inventor attributed, moreover, healing
powers to the touch of the siren, he was soon overrun,
and could, after a short time, retire upon a competency.
An European charlatan purchased the marine monster at

a high price, and exhibited it in England and on the Continent. He met with great success for a time; then he and his siren were forgotten, only, however, to revive more brilliantly than ever in the hands of the master of his art, our own great Barnum. Another siren was, a couple of years ago, the marvel of the rural population all over England; nor was it, in this case, a mere mummy that was shown, but a living mermaid not unattractive in appearance, who discreetly plunged her fish-tail into the waters of a huge basin, and held the classic mirror and comb in her hands. At last public sympathy was aroused by some benevolent Quakers; an investigation was ordered by the authorities, and it was found that the poor woman had been forced for years to spend her days in the water, with an imitation fish-skin sewed on to her body!

There is, however, quite enough that is truly marvellous in some of the greater denizens of the deep, to engage our interest, and to find in them the originals of the fabled beings of whom we have spoken, without resorting to such gross and cruel deception. Pliny already speaks of a sea-elephant, so called at first, no doubt, mainly on account of his two enormous teeth, and of the peculiar shape of his head, which resembles somewhat the trunk of an elephant. A variety of these monsters seems to have been known to the Norwegian Olaus Magnus, who gives a most extraordinary description of the manner in which they were captured. "Sometimes,"

he says, "they fell asleep on the rocky coast, and then
the fishermen went quickly to work, raising the fat
along their tails, and attaching to it strong ropes, which
they fastened to rocks and trees on the shore. Then
they waked up the huge animal by throwing stones at it
with a sling, and compelled it to return into the water,
leaving its skin behind!" At present, the sea-elephant
is found only in the Antarctic Ocean. On the confines of
that world of ice, as far as the eye can reach, there appears nothing in sight but vast masses of ice, thrown in
apparent disorder upon the immense plain, with here
and there a colossal block rising on high and mimicking
the shape of a great palace, with its walls and ramparts,
towers and turrets, battlements and colonnades. Before
these, smaller blocks dance in weird, wearisome motion
up and down on the dusky waters, and gray mists hang
from their sides, and break with their tatters and fragments the dreary prospect. At rare times the sun breaks
through the dense fog, and then the whole world of ice
begins to glitter and glare in the bright rays, and enchanted scenes dazzle the eye. Here is a snow landscape, with hamlets and trees; the larger blocks of ice
resembling snow-covered houses, and the torn and tarnished masses appearing not unlike trees bending under
the weight of hoarfrost, or bushes feathered with light
crystals. The whole enchanted city, with its narrow
canals, is buried in absolute stillness; gulls fly silently
across the clear air, penguins rise and dive again in utter

quiet, and even the sea-elephants lie voiceless, like colos
sal watchdogs, on the steps of the palaces. Only the
low, mournful blowing of a whale, who sends up his airy
fountain of foam, breaks occasionally upon the fearful silence of this magnificent city of ice.

In these inhospitable regions dwells the elephant of
the seas, a monster not unfrequently thirty feet long, and
measuring over sixteen feet in circumference! His powerful teeth are formidable enough in appearance, and
above them he raises, when he is roused to anger, his
inflated trunk, which ordinarily hangs loosely over the
upper lip. His whole body is covered with stiff, shining
hair, and underneath his fur coat he has a layer of fat at
least a foot thick, which protects him effectually against
the terrible cold of the polar regions. The two awkward
feet, mere stumps encased in fin-like coverings, are of
little avail to the giant when he moves on firm land;
after a few yards he begins to groan and to rest, while
the whole huge body shakes as if it were one vast mass
of jelly-like fat. Here he falls an easy victim to the sailors, who come in search of his ivory and his oil; they
walk fearlessly through the thick crowds, and knock
them over by a single blow on the nose. The giant
opens his enormous mouth and shows his formidable
teeth, but, as he cannot move, he is virtually helpless.
Very different are, however, his motions in his own element; as soon as he is under water, he swims with
amazing rapidity, turns and twists like an eel, and is thus

enabled to catch not only swift fish and sepias, but even the web-footed penguins. He must find it difficult, at times, to provide his enormous body with sufficient food, for he swallows masses of tangled sea-tang, and large stones have been found in his stomach, to the number of twelve. When he wishes to sleep, he floats on the surface, and is rocked and cradled by the waves of the ocean.

What has, in all probability, led to their being taken for human beings by credulous and superstitious mariners of early ages, is the beauty of their eye, and the deep feeling they manifest at critical times. They not only never attack men, but, unlike the sympathetic seals, they also abandon their wounded companions, and purposely turn aside so as not to witness their sufferings and their agony. When they are mortally wounded, they drag themselves painfully inland, and hide behind a large rock to die in peace and unseen by others. If they are prevented from thus retiring, they shed tears, as they also weep bitterly when they are ill-treated by cruel sailors.

Very different in temper is the walrus, another of the great monsters of the deep, who, although by nature as gentle and peaceful as the sea-elephant, has become bitter and fierce by his constant warfare with man. It is the true type of the polar North: as all nature here is buried in sad, deathlike silence for several months, so the walrus also sleeps for the same time, deprived of all power and energy, while the fierce tempests and terrible ice-drifts of those regions are represented by their wild passions. They fight with indomitable courage for the fairest among the females, and many a bold knight among them leaves his life in the lists of the grim tournament. They defend their family and their race with intense rage, and know the strength that lies in union. Far up in the coldest ice regions of the Arctic seas they assemble in crowds of two thousand, and when -their guards have been posted, they begin their sports in the half-frozen waters. They splash and splatter as they leap frantically or plunge their huge bodies into the foaming waves, and the noise they thus make, together with the trumpeting of their wide-open nostrils, and the mournful howl of their repulsive voice, fill the air with a stunning, confusing roar. Their appearance is in keeping with the whole scene : black heads, with red, staring eyes of great size, a broad-lipped, swollen mouth, and enormous beard, each hair of the thickness of a straw, adorned with snow-white teeth more than two feet long, and colossal, shapeless bodies, half horse and half whale, but weighing at times not less than three thousand pounds—surely nothing more was needed to strike terror into the hearts of ignorant seamen, and to lead to fancies wild and weird of man-resembling monsters of the deep.

Far greater, however, is the resemblance which certain varieties of seals bear to the human form. Their head, perfectly round and bald, their large bright eyes full of

2*

intelligence and tender feelings, their full beard on both
sides of the face, and their broad shoulders, give to the
upper part of their body a startling likeness, such as, in
the foggy atmosphere of the northern seas, and with a
predisposition to see what people expect to see, may very
well have led to a sincere conviction that they were human beings. To this must be added their merry, playful
disposition, and the peculiar manner in which they hold
themselves almost perfectly upright when gambolling in
the water. Naturally harmless, and even timid, they
have a habit of following the small boats that go on shore,
and of observing attentively all that is done; and if the
crew remains longer at one and the same place, they become familiar, and fond of their company. They learn
to know the people living on the shore near their play-
ground, so that, in Corsica, flocks of them follow the fishing-boats, and modestly content themselves with the fish
rejected after the nets have been hauled in. There can
be little doubt that this intimacy has given rise to an
account, given by Pliny, of a scene daily enacted near
the town of Mines, in southern France ; and as here truth
and fiction meet in striking relation to each other, we
insert the words of the great naturalist:

" At a certain period of the year a prodigious number
of mullets make their way to the sea through the narrow
mouth of a swamp called Latera. These fish choose the
moment of the incoming tide, which prevents the stretching out of nets and the taking them in vast quantities.

By a similar instinct they turn at once toward the open
sea, and hasten to escape from the only place in which
they are liable to be caught. The inhabitants, who know
the period of this migration, and enjoy the pleasure of
the sport, assemble on the shore. Spectators and fishermen all cry aloud: 'Simo! Simo!' Immediately the
dolphins know that they are needed. The north wind
carries the sound of the voice to them. But whatever
time it may be, these faithful allies never fail to appear
at once. One might imagine it was an army, which
instantly takes up its position in the opening where the
action is to take place. They close the outlet to the
mullets, who take fright, and throw themselves into shallow water. Then the fishermen surround them with
their nets. But the mullets, with wonderful agility, leap
over them. Now the dolphins fall upon them, and, content for the moment with having killed them, wait to
devour them when the victory is assured. The action
goes on, and, pressing the enemy closer and closer, the
dolphins allow themselves to be imprisoned with the mullets, and, in order not to frighten them into desperate
acts, they glide stealthily between the boats, the nets,
and the swimming fishermen, so as to leave no passage
open. When all are taken, they devour those they have
killed. But knowing that they have labored hard enough
to deserve more than a single day's wages, they reappear
on the morrow, and not only receive as many fish as they
desire, but are fed with bread soaked in wine!"

The talents of the seal are manifold: from the agility which he displays in catching fish for his master, to the capacity he has shown in learning actually to speak. More than one seal has been taught to utter distinctly the word Papa, and several animals of the kind are reported to have gone even beyond, and to have pronounced several words at a time. Nor must their love of music be forgotten, which is so great that they will rise from the water and remain nearly standing upright as long as the instrument is played, to which they listen with unmistakable pleasure. It is not so very long since one of this remarkable race came every day for six weeks from the waters of the Mediterranean, to take her rest under the divan of a custom-house officer in Smyrna. The latter had tamed her, and placed a few rough planks at the distance of about three feet from the water's edge under his couch, and on these boards the seal loved to rest for several hours, giving vent to her delight, oddly enough, in a profusion of sighs like those of a suffering man. She ate readily the rice and the bread which were offered her, though she seemed to have some trouble in softening the former sufficiently to swallow it with ease. After an absence of several days, the affectionate creature reappeared with a young one under the arm, but a month later she plunged one day, frightened, into the water, and was never seen again.

Nearly about the same time, another seal appeared suddenly in the very midst of the port of Constantinople, undisturbed by the number of caiques dashing to and fro, and the noise of a thousand vessels with their crews and their passengers. One day the boat of the French legation was crossing over to Pera, loaded with wine for the ambassador. A drunken sailor was sitting astride on the cask, and singing boisterously, when all of a sudden the seal raised himself out of the water, seized the sailor with his left arm, and threw himself with his prey back into the waves. He reappeared at some distance, still holding the man under his fin, as if wishing to display his agility, and then sank once more, leaving the frightened, sobered sailor, to make his way back to the boat. Surely nothing more than one such occurrence was needed to give rise to the many romances of former ages; if the same, even, had happened in earlier days, the seal would have been a beautiful Nereid, who, having conceived a passion for the hapless sailor, had risen to take him down to her palace under the waves.

II

ODD FISH.

" And four great beasts came up from the sea, diverse one from another."—
DANIEL vii. 3.

IN the whole range of fabulous monsters, there is not one
that has met with greater incredulity, and yet maintained its hold on the wonder of man with more constant
tenacity, than the kraken. From time immemorial it
has appeared again and again on the pages of travellers,
and from the oldest philosopher to the days of Lacépède
and Buckstone, these faint traces of its true character and
gigantic proportions have been carefully examined, and,
when stripped of the usual exaggerations, been found to
agree with the actual dimensions of a genuine and formidable monster.

Aristotle, whose history has so often been the laugh-
ing-stock of the half-informed, and whom the sceptics of
all ages have been delighted to use as a type of unrelia-
ble naturalists, has of late recovered, step by step, the
veneration which he enjoyed in the Middle Ages. It

would be an interesting task to gather the great facts constantly represented, in scientific works even, as new discoveries, of which a correct sketch is already contained in the works of the ancient savant. Thus he seems to have known, better than any naturalist down to our own day, the nature of the polypus, who in all probability has filled the imagination of men for so many centuries, under the name of the kraken.

Trebius tells us a story, on the other hand, in which undoubted facts are already half-hidden under a mass of exaggerations, of which Aristotle never became guilty, however common they were in the writings of the ancients. A polypus, he says, came every night from the great deep on shore at Carteja, in order to feed upon salt meat. These robberies incensed the people, who in vain tried to discover the intruder, although they surrounded their drying-places with high palisades. The polypus took advantage of a large tree which stood near them, and by means of an overhanging branch that could support his weight, he slipped in night after night. At last, however, his hour came; the dogs discovered him one morning, as he tried to make his way back to the sea, and soon hosts of men surrounded the monster,—at a distance only, for the novelty of the sight, the hideousness of the monster all covered with brine, his enormous size, and the horrible odor which he diffused on all sides, nearly petrified the poor fishermen. In the meantime, he was fighting the dogs bravely, now striking them to the ground with his two larger arms, and now beating them painfully with his whip-like tentacles. At last the men gained courage, and with their tridents they overcame and despatched the monster. We must add, for the honor of Pliny, who quotes the account of Trebius, that he looks upon it as a prodigy, and in his quiet, quaint way, gives the reader to understand his reluctance to vouch for the statement.

The head and the arms of the giant were, however, brought to Lucullus and carefully measured. The former was of the size of a cask, capable of holding fifteen amphoræ, with a beak in proportion; the arms were thirty feet long, and so large that a man could hardly span them; what remained of the flesh, weighed still over seven hundred pounds. Whether Lucullus had it dressed for his tables, is not stated; we know, however, that the Romans were as fond of the flesh of these hideous creatures, as the fishermen of the coasts of Normandy are in our day; it is firm, but savory, and assumes, when cooked, a white and pink color which looks most appetizing.

Fulgosus has a similar story, with such slight variations only, that it appears essentially the same account. Aslian, however, furnishes new evidence; for he states, upon good authority, that a huge monster of the kind, as large as the biggest of whales, was killed with axes by Spanish merchants, whose magazines it was in the habit of robbing. Pliny adds the crowning-piece of wonder:

A polypus, he says, exists in the great ocean, called Arbas, whose feet are of such enormous size that they prevent it from coming into the Mediterranean, as the Straits of Gibraltar are too shallow for such a giant!

Very different are the accounts which take up the thread where antiquity left it suddenly, at the time when Rome fell,—heathen gods were dethroned, and the darkness of the Dark Ages fell like a pall upon all mankind. The Scandinavian seamen, bold like no other sailors on earth, regular vikings, dwellers on the great deep, colored all their relations with the dark and dismal tinges of their grim northern climate. The Greeks and the Romans, even, admired only what was beautiful and graceful in nature, and thus, although they knew the kraken, they loved not to dwell on his monstrous proportions and hideous appearance. Their poetry never alludes to them, and their art disdained to stoop to such repulsive forms. Not so the sombre children of northern twilight : they also know the kraken, and describe it with remarkable correctness in their soberer moments; but they love to dwell upon its repulsive features; they exaggerate its dimensions and its ugliness; they change it into a terrible being, full of dread power and malign purposes; and then they believe in their own dreams and enjoy like children the strange delight with which they are filled by their very fears. They go on increasing its size, till it becomes, to their excited imagination, the Mountain Fish, and they see it soon everywhere, in their land-locked bays and

out on the stormy sea; when the thick storm-clouds lower
till they touch the crest of the waves, it is the kraken;
and when their anchor suddenly strikes upon an unknown shallow, it is again the kraken.

The Norwegians, especially, loved to tell wondrous
tales; how their bold seamen landed on a deserted island
which showed no trace of life, not a shrub nor a blade of
grass, and while they still wandered about, marvelling
at the utter desolation, the island began to heave and to
move, and behold! they found themselves on the back
of the monster! Great authorities came to confirm the
stories; saints and bishops lent the weight of their sacred
character to the accounts given by laymen and heretics.
Erick Falkendorf, a bishop of Nidros, wrote, in 1520, a
long letter on the subject to Pope Leo X. He was sailing, on a Sunday, in a Norwegian vessel along the distant coast, and bewailed his inability to celebrate holy
mass on firm land. As he mourned and prayed, suddenly an unknown islet arose, not far from the vessel;
the crew land, the sacred vessels are carried on shore,
and the holy office is celebrated with due solemnity.
After mass they return on board ship, and immediately
the island begins to tremble, and gradually to sink back
into the sea, from which it had risen. The island had
been a kraken!

Olaus Wormius, also, who is generally truthful enough,
relates having seen, about the year 1643, one of these
enormous monsters, and states that they resemble an

island far more than an animal. He expresses his belief that there are but few krakens in existence, and curiously enough adds, that while they are themselves immortal, the Medusæ are nothing more than the eggs and the spawn of these monsters.

Other writers, of the same century, confirm his statement, and believe in the immortality of the kraken—a faith which was not even shaken when, in 1680, the carcass of one of these monsters was for the first time discovered in the Gulf of Newangen, in the parish of Astabough. His arms had become entangled in the countless cliffs and rocks which characterize the dangerous coast of that neighborhood, and the animal had died there, unable to extricate itself. When putrefaction commenced in the enormous mass, the odor became so offensive for miles and miles, that serious fears of a pestilence were entertained. Fortunately, the waves came to the aid of the frightened people, tearing off piece after piece, and carrying it into the ocean; and when the last remnant had been washed away, an official report of the whole event was drawn up by a clerical dignitary, and is still to be found in the government archives at Drontheim.

A similar case occurred on the Newfoundland banks, where polypi abound in such numbers that the fishermen of all nations, who congregate there in the season, use, every summer, nearly two millions as bait, with which to catch codfish. Towards the end of the last century, a monstrous specimen of this class died on these

banks, beyond Pine Light, and here, also, the mass of pu-
trifying matter was so enormous, and the odor so intoler-
able for a great distance, that the grave apprehension of
an epidemic drove the fishermen from the neighborhood,
till the currents had carried off every trace of the terrible animal.

Of all the authors, however, who have given us more
or less detailed accounts of their experience with the soe-
trolden, or sea-scourge, as the Swedes call it, Pontoppidan
is by far the most precise in his statements.

The northern people, he tells us, assert, and without
the slightest contradiction in all their accounts, that when
they go out into the open sea, during the great heat of
the summer, they find the water suddenly less deep, and
upon sounding, the lead frequently marks only thirty
fathoms. The fishermen know then that a kraken floats
between the lead and the bottom of the sea, and they
immediately get ready their lines, for they know that
where the monster is, fish always abound. If, on the
other hand, the depth diminishes, if this accidental bottom moves and rises, then it is time for them to make
their escape ; for the kraken is waking up and about to
rise, in order to breathe and to stretch out its huge arms
towards the sun.

The fishermen hasten away with all their might, and
when they can at last rest on their oars at a safe distance,
they then see the enormous creature, whose back covers
a mile and a half of sea. The fish, taken by surprise by

his sudden rising, leap frantically about in the small pools formed in the rugged irregularities of his back, and then a number of points or shining horns appear, gradually rising till they look like masts with their yards ; these are the arms of the kraken, which are so powerful that they can seize the ropes of a large ship and sink it in a few moments. After having remained a short time above water, the kraken sinks down again, and this is scarcely less dangerous to vessels near by, as he displaces, in sinking, such an enormous volume of water, that whirlpools and currents are formed, scarcely inferior to those of the dread maelstrom.

Such is the account found in the Natural History of the learned bishop, who, no doubt, wrote what he conscientiously believed to be true, although he cannot quite disguise his own scepticism in regard to some of the facts mentioned. It is very different with Augustus, of Bergen, a man of critical mind, who, not having seen a kraken himself, collected all the Scandinavian accounts of which he heard, and, after examining them carefully, came to the conclusion that there does exist a gigantic polypus—though far from boasting of the dimensions usually attributed to the monster—that it is provided with arms, that it emits a strong odor, that it shows at times long ventacles, and only appears in summer time, and during calm weather. It is remarkable how fully the conclusions of this learned naturalist have been confirmed by modern discoveries.

The great Linné, a Swede in heart as in race, seems to have been troubled with strange doubt concerning this pet monster of his countrymen; for, after having solemnly introduced the kraken into his Swedish fauna, and after speaking of it even more fully in his great work, "The System of Nature," he suddenly drops him in the seventh edition, and never more says a word of the gigantic polypus. This did not have much effect, however, on the sailors of his and of other lands, as they were not much given to reading Latin works; and in Sweden, as well as in France, the faith in the kraken remained as general and as firm as before. Countless votive offerings adorn, to this day, the little chapels that rise high above the iron-bound coast, with their tiny turrets and tinkling bells; but none more weighty in precious metals, none more thankfully offered to the Lord of the Sea, than those which speak of the delivery from the dread kraken. One of these—in the church of "Our Lady of the Watch at Marseille"—is accompanied by a touching recital of a fearful combat with the monster on the coast of South Carolina, and another, hung up in the chapel of St. Thomas, at St. Malo, testifies to the escape of a slave-ship from the arms of a gigantic polypus, at the very moment when it was leaving the port of Angola.

In 1783, a whaler assured Dr. Swediaur that he had found in the mouth of a whale a tentacle of twenty-seven feet in length. The report was inserted in a scientific journal of the day, and there read by Deny Montfort, who at once determined to obtain more ample information on the subject. It so happened that, just then, the French Government had sent for a number of American whalers, in order to consult with them as to the best means by which the French fisheries could be revived. These men were staying at Dunkirk; and here Montfort questioned them, and upon inquiry it appeared that two of them had found feelers, or horns, of such monstrous animals. Ben Jonson saw one in the mouth of a whale, from which it hung to the length of thirty-five feet; and Reynolds another, floating on the surface of the sea, forty-five feet long, and of reddish slate color. But of all the reports which he heard, the following was the most minute and yet also the most extraordinary:

Captain John Magnus Dens, a Danish sailor of high character and established uprightness, deposed that, after having made several voyages to China in the service of the Gottenburg Company, he had once found himself becalmed in the fifteenth degree S. L., at some distance from the coast of Africa, abreast of St. Helena and Cape Nigra. Taking advantage of his forced inactivity, he had determined to have his ship cleaned and scrubbed thoroughly, and, for that purpose, a few planks were suspended on the side of the vessel, on which the sailors could stand while scraping and caulking the ship. They were busy with their work, when suddenly an anchertroll —so the Danes call the animal—rose from the sea, threw one of its arms around two of the men, tore them with a jerk from the scaffolding, and sank out of sight in a moment. Another feeler appeared, however, and tried to grasp a sailor who was in the act of ascending the mast; fortunately, the man could hold on to the rigging, and as the long feeler became entangled in the ropes, he was enabled to escape, though not without uttering most fearful cries. These brought the whole crew to his assistance; they quickly snatched up harpoons, cutlasses, and whatever they could lay hands on, and threw them at the body of the animal, while others set to work cutting the gigantic feeler to pieces, and carrying the poor man to his berth, who had swooned from intense fright. The monster, with five harpoons thrust deep into its quivering flesh, and holding the two men still in its huge arms, endeavored to sink; but the crew, encouraged by their captain, did their utmost to hold on to the lines to which the harpoons were fastened. Their strength was, however, not sufficient to struggle with the marine giant, and all they could do was to make fast the lines to the ship, and to wait till the forces of the enemy should be exhausted. Four of the ropes snapped, one after the other, like mere threads, and then the harpoon of the fifth tore out of the body of the monster with such violence that the ship was shaken from end to end; thus the animal escaped, with its two victims. The whole crew remained overcome with amazement; they had heard of these monsters; but never believed in their existence; and here, before their eyes, two of their comrades had been

torn from their side, and the third, overcome with fright,
died the same night in delirium. The feeler which had
been separated from the body, remained on board as an
evidence that the whole had not been a frightful dream;
it measured at the base as many inches as their mizzenmast, was still twenty-five feet long, and at the pointed
end provided with a number of suckers, each as large as
a spoon. Its full size must have been far greater, however, as only part of it had been cut off, the animal never
even raising its head above the surface. The captain,
who had witnessed the whole scene and himself thrown
one of the harpoons, ever afterwards considered this
encounter the most remarkable event of his checkered
life, and calmly asserted the existence of the kraken.

All these ancient accounts, the Norwegian legends, the
reports of sailors of many nations, and the minute descriptions of Sicilian divers, who spoke of polypi as large
as themselves, and with feelers at least ten feet long,
could not fail to make an impression upon men of science,
and the most discreet among them came to the conclusion
that there must be some truth amid all those fables.

It was, however, reserved to our century to strip the
facts of all exaggeration, and to establish the existence of
such monsters beyond all controversy. A kind of mollusk, called cephalopodes, were found in various seas,
whose peculiar formation and strange appearance sufficiently explained the marvels told of the kraken. An
elongated sack in the form of an egg, or a cylinder, from

3

which protrudes at one end a thick, round head, with a pair of enormous flat eyes; on this head, at the summit, a kind of hard brown beak, after the manner of a parrot's bill, and around the beak a crown of eight or ten powerful, long arms—this is the polypus, which passed of old for a kraken.

On the inner side, each one of these gigantic arms or feelers is covered with a double row of suckers, which resemble a small cup with a movable bottom. By means of these cups, which the animal can exhaust of the air they contain, it can affix itself to any surface; and as it possesses several hundred of them, its power is naturally enormous. Nevertheless, they use these feelers only for the purpose of seizing their prey and handing it up to the beak, which then goes to work and tears it to pieces. Nor is their manner of swimming less curious. Their gills require a large quantity of water to furnish them with a few globules of air; to provide this supply, they are covered with an elastic mantle, which the animal contracts when it is full, so as to drive the water it contains through a tube placed between the eyes. Every time that the mantle is thus contracted and the water expelled, the latter forms a kind of jet, which, striking upon the inert matter around, gives to the animal an impetus in the opposite direction. At each pulsation, therefore, it advances, and thus it travels quite rapidly through the water.

The cephalopodes on European and American coasts

are generally only of small size, although in the Mediterranean and the Adriatic seas some have been found of
larger dimensions, and others still greater are kept in
museums. In the open sea, however, vessels have encountered genuine giants of the kind, and these are, no
doubt, the true representatives of the kraken. Rang
met one of the size of a ton, and of reddish color, while
Pennant saw in the Indian Seas an eight-armed cuttle-
fish, with arms of fifty-four feet in length and a body of
twelve feet in breadth ; thus making it extend, from point
to point, one hundred and twenty feet. A naturalist of
Copenhagen, who has made the study of these animals
his specialty—Steenstrup—had occasion to examine one
of these monsters in 1855, on the coast of Gothland,
where it had been caught by fishermen. It required several carts to carry the body off; and the hind part of
the mouth, which he saved from destruction, still had
the size of an infant's head. The museum at Utrecht
contains a specimen of a colossal cephalopode ; and our
own Mr. Barnum was, of course, not without at least
one of the feelers of such a giant, as thick as a man's
body, and over ten feet long. Wm. Buckland, the
great naturalist's son, and an excellent observer himself,
took pains to examine the varieties known to the British
coast, and allowed one small specimen to grasp his hand
and arm. He describes the feeling to be such as if a
hundred tiny air pumps had been applied at once, and
little red marks were left on the skin where the suckers

had been at work. " The sensation," he says, " of being held fast by a (literally) cold-blooded, soulless, pitiless and voracious sea-monster, almost makes one's blood run cold. I can now easily understand why they are called man-suckers; only the natives of the Chinese and Indian seas have such a horror of them; for in those climates they are seen large and formidable enough to be dangerous to any human being who may be so unfortunate as to be clutched by them." Victor Hugo's description of his monster in the "Travailleurs de la Mer," is, of course, far more graphic and poetical, though hardly less to the point. On the logs of many a vessel, encounters with colossal mollusks of this kind have since been entered, and although the largest ever accurately measured—by a French man-of-war's men—was only twelve feet long in body, with feelers of fifty feet length, enough has been seen and recorded in our days to justify the conviction that the wonders of the deep are not yet all known, and that animals like the kraken may be still in existence.

The twin brother of the kraken, both in its marvellous size and in the incredulity which all descriptions have excited, is the famous sea-serpent. Its history is as old as the oldest record; no age and no seafaring nation has been without some account concerning its appearance, and yet to this day serious doubts are entertained as to its existence. It is clearly referred to in the Old Testament, where the prophet Isaiah sings:

In that day the Lord with his sore and great and strong sword
Shall punish leviathan, the piercing serpent,
Even leviathan, that crooked serpent,
And he shall slay the dragon that is in the sea.

Again, when Job plead his uprightness, and the Lord answered unto Job out of the whirlwind, he mentions behemoth and leviathan, and says concerning that monster:

Canst thou draw out leviathan with a hook?
Or his tongue with a cord which thou lettest down?
Canst thou put a hook into his nose or bore his jaw through with a thorn?
   *   *   *   *   *
Shall thy companions make a banquet of him?
Shall they part him among the merchants?
Canst thou fill his skin with barbed irons? or his head with fish spears?
   *   *   *   *   *
Who can open the doors of his face? His teeth are terrible round about.
His scales are his pride, shut up together as with a close seal.
One is so near to another, that no air can come between them.
By his neesings a light doth shine, and his eyes are like the eyelids of the morning.
Out of his mouth go burning lamps, and sparks of fire leap out.
Out of his nostrils goeth smoke as out of a seething pot.
His breath kindleth coals, and a flame goeth out of his mouth.
   *   *   *   *   *
When he raiseth up himself, the mighty are afraid:
The sword of him that layeth at him cannot hold: the spear, the dart, nor the habergeon.
He maketh the deep to boil like a pot: he maketh the sea like a pot of ointment.
He maketh a path to shine after him; one would think the deep to be hoary.
Upon the earth there is not his like.

It is well known that the monster, so graphically described by the voice that spoke to Job in the whirlwind, has been sometimes believed to be the elephant, and then again the crocodile. But in the Psalms, Leviathan is distinctly mentioned as living in the great and wide sea, and God is said to have formed him to play therein.

The Jews, therefore, evidently looked upon him as a sea-serpent of colossal size and most formidable appearance, identical with the instrument of the Lord, of which He says: "And though they be hid from my sight in the bottom of the sea, thence will I command the serpent and he shall bite them."

These imposing descriptions are, moreover, by no means limited to the excited imagination of Hebrew writers; other nations also record in their annals the existence of such a gigantic wonder of the deep. Palladius, for instance, speaks of a serpent of the Ganges, which he calls grandly an odontotyrannus, who could swallow an elephant without straining. Solin has heard of him frequently, and knows that he lives in India and Ethiopia, crosses the Indian Ocean by swimming, and travels from island to island; while Pliny quotes Solam, who found these colossal serpents in the Ganges; "They were blue," he says, "and so large that they could easily seize and drag under water an elephant."

It is, however, in the Middle Ages, that we find the fullest accounts of the monster. Pontoppidan, one of the most learned Scandinavians, who was long bishop of Bergen, in Norway, and died as chancellor of Denmark, in Copenhagen, in 1764, states, in his interesting contributions to Natural History, that in his country every body believed firmly in the great sea-serpent, and if he or any of his guests ventured to speak doubtingly of the huge monster, all smiled, as if he had been uncertain whether

eels or herrings really existed. The good people of those
northern regions were so familiar with these wonderful
creatures, that they spoke of two distinct kinds of sea-
serpents, one living in the sea only, the other amphibious,
which preferred the land generally, but retired periodically to the great deep. Nicolaus Gramius, minister of
the gospel at London, tells us, that during a destructive
inundation, an immense serpent was seen to make its
way towards the ocean, overthrowing every thing in its
path, animals, trees, and houses, and uttering fearful roar-
ings. The fishermen of Odal were so frightened by the
terrible sight, that they did not dare go out in their
boats for several days.

The famous Archbishop of Upsala, Olaus Magnus,
who bears testimony to the kraken, also speaks more
than once of the amphibious serpent. He states that
they leave the shelter of the cliffs near Bergen at night;
they have a mane, their bodies are covered with scales,
and their eyes send forth a bright light; out at sea they
rear themselves against the ship they encounter, and seize
whatever they can obtain on deck. An animal of this
kind, he continues, was actually seen in 1522, near the
island of Moos, which measured over fifty feet in length,
and was continually turning round. Several works on
Natural History, down to the celebrated compilation of
H. Ruysch, published in 1718, contained pictures of
these Norwegian serpents. Paul Egede, finally, a most
trustworthy man, and well known by his connection

with Greenland missions, which he helped his father to establish and fostered as bishop, not only bears witness to the frequent appearance of such sea-serpents on the Scandinavian coasts, but describes minutely one which he met himself on his second voyage to Greenland.

If we believe, therefore, the testimony of the Hebrews and of the Northern nations, there exists a serpent, living in the sea, of gigantic proportions, swimming by vertical movements, in which it is aided by fins which hang down from behind its neck, as was the case in fossil reptiles like the plesiosaurus, and covered with a thick skin, which was frequently found cast off on desert islands. On the back it has a shaggy mane, its eyes are large and brilliant, and the head is shaped somewhat like that of a horse. It is only seen in midsummer and during fine weather, for the want of stability in its long flexible body disables it from resisting the effect of high winds.

Like all rare animals of gigantic proportions, the sea-serpent naturally excited terror in the few persons who ever encountered it; and this led, as a matter of course, to marvellous stories about its size and ferocity. Sailors loved to tell how the great monster would throw itself bodily over their vessel to make it sink, and then to feed at leisure on the bodies of drowned seamen. Others told frightful tales of beloved comrades, who were suddenly snatched from their side by such animals, which appeared unexpectedly at the ship's side, raised their colossal head and neck, and instantly disappeared again with their victims. Fortunately, a very simple means of defence is at hand, according to the belief of Norwegian sailors; these monsters have, it seems, a most delicate sense of smell, and cannot endure the odor of musk; nothing is needed, therefore, but to scatter some musk on deck, and the terrible animal no sooner smells it from afar, than it makes off and dives to the very bottom of the sea.

It was, however, by no means in remote times only, that the sea-serpent has been seen by travellers and sailors. Reports, on the contrary, abound in our day also of such encounters, and scarcely a season passes but the monster has been seen in our own seas, or in more distant parts of the northern ocean. Officers of the navy, ministers of the gospel, American whalers and English navigators, all agree, without essential differences in their statements, on the existence and general forms of such monsters.

The first reliable account of this great enigma of modern days came from a ship captain, Laurent de Ferry, of Bergen, in the form of a letter, from which we extract the following facts: Towards the end of the month of May, in 1746, he was returning from a voyage to Trundhin, when the weather being calm and warm, he suddenly heard the eight men, who formed the crew of his boat, whisper to each other. He laid aside the book which he

3*

was reading, and then noticed that the man at the helm
was keeping off from the land. Upon questioning the
latter, he was told that there was a sea-serpent right
before them. He ordered the man immediately to turn,
and to steer straight upon the strange animal, of which
he had heard much during all his life. The sailors at
first showed great reluctance to obey, but as the monster
was right before them, and moving in the same direction,
they became excited, and after a while engaged heartily
in the novel stern chase. The captain, fearful that the
monster might escape, fired his gun at it, and immediately
it plunged, evidently wounded, for the water all around
was stained red, and remained so for two or three minutes. The head, which rose over two feet above the
surface, resembled a horse's head : it was gray, with a
dark-brown mouth, black eyes, and apparently a long
mane floating over the neck. Beyond the head they
could see seven or eight coils of the huge serpent, each
of enormous size, and at considerable distance from the
next. The animal did not reappear; but the time during which it was clearly in sight was ample to enable
the captain and his crew to examine it closely.

The only other report which is perfectly clear and precise, has a Rev. Mr. McLean for its author, who wrote
from the Hebrides, and very naïvely exhibited his terror.
His statement amounts to this : He saw the sea-serpent
in June, 1808, on the coast of Coll. He was sailing about
in a boat, when he noticed, at the distance of half a mile,

an object which excited his surprise more and more. At
first he took it for a small rock among the breakers; but
knowing the sea very well, and being sure that there was
no rock there, he examined it carefully. He then saw
that it rose considerably above the surface, and after a
slow, undulating movement, he discovered one of the
eyes. Alarmed at the extraordinary appearance and the
enormous size of the animal, he cautiously coasted along
near the land, when he suddenly saw the creature plunge
in his direction. He as well as his men were thoroughly
frightened, and pulled with all their might to escape.
At the very moment at which they reached the shore,
and when they had barely had time to climb up to the
top of a large rock, they saw the monster glide slowly up
to their boat. Finding the water quite shallow there, it
raised its horrible head, and turning again and again,
seemed to be troubled how to get out of the creek. It
was seen for half a mile, slowly making its way out to
the open sea. The head was large, of oval shape, and
rested on a rather slender neck. The shoulders, as the
good pastor calls them, were without gills, and the body
tapered off toward the tail, which was never distinctly
seen, as it was generally under water. The animal
seemed to move by progressive undulations, up and down;
its length they estimated at from seventy to eighty feet;
it moved more slowly when the head was out of water,
and yet it raised it frequently for the evident purpose of
discerning distant objects. At the same time when Mr.

McLean saw the serpent, it was also seen in the waters near the Island of Carma. The crews of thirteen fishing-boats were so frightened by its terrible appearance, that they sought refuge in the nearest creek.

Nor were they only seen out at sea when mistakes would be natural, and fright or intense curiosity might lead to unconscious exaggeration, but the body itself has been found and examined by competent persons. Thus, to mention but one instance, in 1808, the body of a gigantic serpent was washed on shore at Stronsa, one of the Orkneys. A Dr. Barclay was summoned at once, and, in the presence of several justices of the peace and some men of learning, an affidavit was drawn up, which stated that the monster measured over fifty feet in length and nine feet in circumference; that it had a kind of mane running from behind the head nearly to the tail, which was brilliantly phosphorescent at night; and that its gills, nearly five feet long, were not unlike the plucked wings of a goose. Sir Everard Home, it is true, believed it to be a basking shark of uncommon size. But Gloucester fishermen repeatedly saw similar animals, and the Linnæan Society of the United States examined carefully a number of witnesses; the same has been done in Holland and in the Dutch colonies of Java, and everywhere evidence has been obtained, which showed remarkable unanimity, and precluded the idea of such a mistake.

Does the sea-serpent belong only to the realm of fancy, or is it really one of the great wonders of the deep? The question has never yet been finally decided. That there must be in existence animals of serpent-like form and of gigantic proportions, seems to be well established by the concurrent testimony of American, English, and Norwegian eye-witnesses; and the mere fact that no specimen exists in the museums, and that no such monster has been encountered of late years, does not by any means disprove the experience of so many ages. On the other hand, great allowance must no doubt be made for the effect of fear which enlarges all objects, the desire to excite wonder which leads men to embellish their accounts, and the natural tendency to add to original accounts, which results often unconsciously in exaggeration, and has in all probability furnished us with such wonderful creatures as the kraken, the roc, and the phœnix.

Nothing in these descriptions is, besides, actually incompatible with the laws of nature; and the study of fossil remains establishes beyond doubt the fact, that in former ages gigantic reptiles have peopled the sea, which were far more surprising in size and shape than the much-doubted sea-serpent. There is no necessity, therefore, to ascribe all such encounters to simple mistakes; now and then, perhaps, a long string of algæ, moving slowly under the impulse received by gentle winds or unknown currents, or masses of phosphorescent infusoria, floating for miles and miles on the calm surface of the sea, may have led superstitious sailors to fancy they saw giant serpents. But it is, on the other hand, by no means im-

probable that the vast deep, of which so little is as yet
known to man, may still hold some of the giants of olden
days, and that, of the many well-trained, intelligent people who nowadays " go down to the sea in ships, and do
business in great waters," some may yet see these " works
of the Lord, and His wonders in the deep." Fortunately,
there is no doubt any longer surrounding the true king
of the seas—the whale—and yet, he is of truly monstrous
proportions. It is a perplexed chapter, to be sure, in
natural history, to say how many species of whales there
are ; for Jack Tar comes home with confused accounts
of Sulphur Bottoms, Broad Noses, Razor Backs, and Tall
Spouts, and a host of other names by which he learns to
distinguish unprofitable whales, not worth the toils and
perils of capture. But after all, this only shows that the
family is very fully known ; and well might this be so,
if we remember that already in the ninth century a Norwegian Ohthere, whose wonderful adventures were taken
down by no less a man than King Alfred himself, speaks
of having slain sixty such monsters in two days. This
is, of course, impossible ; but we must bear in mind that
there is nothing in nature so wonderful that the human
mind does not love to add a finishing touch of its own,
and make it yet a little more monstrous. Thus, the size of
the largest of moving things, by whose side even elephants
are but dwarfs, has been exaggerated ; and great authorities, of recent times even, have gravely described it as
two or three hundred feet long. Nor must we forget,

that since the Biscayans and Basques first dared attack the whale on the high seas, in 1575, all seafaring nations have joined in the pursuit, and naturally tried to outstrip their rivals in startling accounts of the prey which they hunt on the hungry waves, with toppling icebergs around them.

In early times the Esquimaux killed the true whale of the North with harpoons, to which large bladders of air were fastened, which prevented the poor animal from sinking and plunging; and in our own day, good-sized steamers go out on the whale fishery and dispatch them by the aid of galvanic batteries. Their homes, also, have changed with the times; the Floridians, who were once reported to kill them by driving pegs into their blow-holes, no longer see them on their shores; while new varieties, formerly neglected on account of their activity and energy in self-defence, are now eagerly sought after in distant seas. In olden times, however, strange stories were current about the peculiarities of whales, and among them the accounts of men swallowed by them hold a prominent place. They arose, no doubt, from the biblical account of Jonah, who, "was three days in the whale's belly;" but as, unfortunately, the animal is so made that the mouth will easily hold a boat and its crew in its vast space, but the throat is too narrow to allow even a mackerel to pass, the "great fish" must have been either another variety, or, as some think, a vessel called by its name. Nevertheless a great author, Fournier, recites gravely in his Hydrography, the following story :

During the reign of Philip II., King of Spain, there appeared in the great ocean a whale, very different from all others in this, that he floated partly above the surface, and had large wings, by means of which he could move like a vessel. A ship encountered him, and broke one of these wings by a cannon-shot, whereupon the whale entered, very stiffly, through the Straits of Gibraltar, and uttering horrible bellowings, went ashore near Valencia, where he was found dead. The skull was so enormous that seven men could stand in it, while the palate would hold a man on horseback; two dead men were found in the stomach; and the jawbone, seventeen feet long, is still to be seen in the Escurial.

Nor is this the only fable that has been from of old connected with these true giants of the sea; their size especially has given rise to amusing exaggerations; and the love of the marvellous, which tempts us all, has caused the most extraordinary stories to be spread far and wide, while the more exact, but less attractive descriptions, have been culpably neglected.

Thus Pliny already gravely informs us that there were in his times whales found in the Indian Ocean which measured nine hundred feet; they would, therefore, have easily filled a whole village. Not only romance-writers, but even naturalists of renown, like Gesner, in his work on Fishes (1551), have taken pleasure in representing whales as animals resembling islands; and in speaking of sailors who had landed unawares on their back, covered, as it was, with a mass of green algæ. Saint Ambrosius, and other saints, came near being lost by such an error, if we believe the legends of the Church. The same amusing idea occurs in that delightful book, the "Arabian Nights," which, it is well known, was compiled from very ancient Arabic legends and manuscripts. "One day," says Sinbad the sailor, "as we were sailing along, a calm befel us near a little island. The captain ordered the sails to be furled, and allowed all who chose to go on land; I was among those who landed. But whilst we were amusing ourselves, eating and drinking, the island suddenly trembled, and gave us all a violent shock. It was a whale."

The fable spread rapidly, especially among nations who lived far from the sea and had no means to verify the truth of such accounts by their own experience. They were all the more readily believed, as for generations no other books were accessible to the masses but the Bible and Pliny; and both of these authorities spoke confidently of these monsters; the latter by name, the former, as was then universally believed, under the thin disguise of the leviathan. In the Orient, of course, greater exaggerations still were added, age after age, such as the utter ignorance of the seas prevailing among Eastern nations, and their high-wrought imagination, loved to invent and to hear. An ancient Jewish work, the Bara-

Bathra, already speaks of a vessel which sailed three days
over a whale, before it accomplished the distance between head and tail : and Arab authors loved to assert
that the earth was actually resting upon a gigantic whale
whose occasional tremors were the cause of earthquakes.
One day, they add, the Evil One approached the animal,
and, laughing at the patience with which it had so long
borne an useless burden, persuaded it to shake its back-
bone, and thus to rid itself of the load. The globe was
just about to be shaken off, when, fortunately, some one
informed Allah of the impending calamity, who hastened
to the spot, and, after a long discussion, succeeded in ex-
acting a promise that the whale would continue to uphold
the earth a few thousand years longer. The Chinese are,
as usually, not behind other nations in early and magnificent accounts of their own giants. In an ancient work
of authority, called Tsi-hiai, they speak of a whale Pheg,
which beats four hundred miles of ocean into foam whenever it moves. At a very advanced age, this monster
of the deep is changed ; it becomes then the famous monster of the air, the bird roc.

Now, as we have stated before, the natural history of
the whale is perfectly well known, and we will, therefore,
here mention only one feature connected with the giant
which is not as familiar to all. This is the fact that the
flesh of the whale is excellent food, and was formerly esteemed most highly. It was for a long time a royal dish
in England, and, in 1243, Henry III. summoned the sher-

iffs of London to furnish him, for his table, with a hundred whales! In the thirteenth century it reappears in the housekeeping-book of the Countess of Leicester, and for several generations afterwards all the whales caught in the Thames belonged by right to the Lord Mayor of London, who had them served up in state at great municipal dinners. Marteus, in his northern voyages, ate whale flesh frequently; but he considered it coarse and tough, inferior to beef; the tail only furnished, when well cooked, really palatable dishes. The Normans used to be, in former times, the caterers of the English; they possessed the secret of several recipes by which to cook the rare delicacy, and generally served the pieces of meat with tender green peas.

The main use to which the whale is put, remains, however, the oil, the sperm, and the whalebone, and these are rich sources of income to many nations. American sperm whaling, especially in the Pacific, has, in a commercial point of view, grown to immense proportions. When New England was yet a colony, gray-headed men would point to the sea, saying, " Those are the fields where our children will reap their harvests; " and a fleet of over six hundred sail yearly employed in hunting the sperm whale now realizes the prediction.

Among the near kindred of the whale is the famous unicorn—not the companion of the lion on the British coat-of-arms, but its counterpart among the wonders of the deep. Inferior in size to the right whale, it has the

advantage of a most formidable weapon, with which
Nature has provided it for as yet unknown purposes.
This is the monstrous tooth which projects from the upper jaw of the animal; it is as large as a man's thigh at
the base, turned in a spiral, and sharply pointed at the
end; hollow within, it shows externally the finest and
whitest ivory known to the trade. The Narwhal, or
nose-whale, was so called because the Dutch, who seem
to have had the christening of most quaint things in
northern regions, at first took this horn, projecting
straight ahead, ten or even fifteen feet, for a grotesque
long nose. Some say the animal uses this odd append-
age to pierce holes through the ice when he comes up to
blow or breathe; others, that he mows off seaweed with
it, on which he grazes. There is no doubt that, at times,
he transfixes fish with his gigantic stiletto, so that he
may be able to devour them at leisure.

The legend has it, that a king of Denmark, wishing to
make somebody a present of a piece of the horn of the
unicorn—for such it was long considered—ordered one
of his high officials to cut off a piece at the thicker end
of a fine specimen which he possessed. The officer did
so, and, to his astonishment found that what he had
looked upon as a solid horn, was hollow, and in the con-
cavity he discovered a smaller horn of the same shape
and the same substance. The latter was about a foot
long, and this resemblance to the teeth of men first led,
it is thought, to the idea, that the unicorn might after all

be nothing more than a gigantic tooth. In those days, however, the superstitious people attached marvellous powers to the wonderful horn, and a brisk trade was carried on in fine specimens, and even in broken fragments.

The male alone possesses this formidable weapon; the female having, instead, two small teeth, of little use for purposes of attack or defence. In the male, however, one of these two is disproportionately developed, while the other remains either of diminutive size, or disappears gradually altogether; very much as is the case with the claws of certain crustaceæ. At first sight, it would appear as if this giant of the seas, with his terrible sword, would be the terror of the seas, killing and devouring all that came near him. In reality, however, the narwhal is a very harmless animal, and generally his own enem more than that of others. His mouth has no teeth, and immovable lips, and is so small that he can swallow little else but mollusks and little fish; and Scoresby, who found in the stomach of one of these strange beings a ray of two feet length, came to the conclusion that the fish must have been first transfixed by the tooth, and killed before it was devoured. Otherwise it would have been difficult to understand how an active fish should have allowed itself to be caught by an animal unable to seize it with the lips or retain it with the tongue, and in a mouth which had not even teeth to tear it to pieces.

Their swiftness, when they are alone, is marvellous;

and their capture would be almost impossible, if it were not for the curious habit they have of travelling in immense troops, and of taking refuge in little bays, from which they cannot easily escape. Small boats approach them, in such cases, with precaution; the poor animals begin to crowd upon each other, they press their ranks so closely that soon their movements are impeded, and their enormous weapons become interlaced, as each one tries to raise the head high into the air. They can neither escape nor defend themselves, and thus fall an easy prey to the lances of the whalemen.

Scoresby thus once encountered, on his voyage to Greenland, a troop of narwhals, divided into smaller bands of fifteen or twenty. The males were far more numerous than the females. They seemed to be full of sportive gayety, raising their huge weapons high above the water, crossing them with each other, and uttering a sound as if they were gurgling water in their throats, while they seemed to amuse themselves with the play of the rudder in the water. At other times, however, they are known to be in a very different humor, and then they attack and sometimes pierce large whales. It is doubtful whether their efforts against vessels arise from illhumor and pugnacity only, or from an idea that the ships are large whales. Like the bees, the poor narwhals also generally seal their own doom when they make such attacks; for the enormous tooth, driven with prodigious force into the timber, remains fast there, and, breaking off, causes the death of the ferocious^a animal. At times, when he has driven it in right at the stern, the poor creature itself is fastened to the ship and towed along, until it dies and decomposes, to the great disgust of the sailors, who see their course impeded and their senses insulted without any profit.

In the Paris Museum there is a complete skeleton of a magnificent narwhal, with a tooth of amazing size. The marine monster here shows its exquisite adaptedness to the element for which the hand of the Creator had fashioned it, and no one, on seeing the slender, flexible form, can doubt its far-famed agility and terrible strength.

The Greenlanders eat the flesh, and obtain from the fat an oil second only to the best sperm oil. But it is the tooth, after all, which has made the narwhal, at all times, one of the wonders of the deep. Long before the animal itself was known, the tooth was familiar to traders as the horn of the unicorn. The monastery of St. Deny's possessed a pair of these remarkable weapons, famous for their size and the beauty of the ivory; they are now in the Medical Museum of Paris. A larger one, nearly nine feet long, exists in the treasury of the Danish monarch, at Fredericksborg.

When they were not kept thus, as most rare curiosities—the unicorn itself having, of course, never been seen—they were manufactured into weapons of every kind, swords and daggers. But they were also endowed, in popular belief, with a wondrous power of counteracting all poisons; and their mere presence, it was thought, sufficed to defeat any attempt at poisoning the owner. Down to the days of the French king, Charles IX., a piece of the precious substance was regularly dipped into the cup of the monarch before he drank; and when the great founder of modern surgery, Ambroise Paré, was requested to raise his voice against the superstition, he replied that the belief was universal; and if he ventured to contradict it, he would be treated like an owl appearing in bright daylight, which the other birds fall upon and kill, and then think no more of the murdered victim. Nevertheless, he subsequently wrote openly against the custom, and with so much skill and power, that after that time no one dared avow his secret faith in the virtue of the unicorn's horn as an antidote.

Wormius, whom we have mentioned before, was the first to establish the true character of the strange curiosity. "Finding myself," he writes, "a few years ago, at the house of Mr. Fris, Grand Chancellor of Denmark, I complained of the want of curiosity in our Greenland merchants, that they should never have inquired after the animal from whom these horns were obtained, or brought home a part of their skin at least. They are more curious than you think, replied the Chancellor, and let me see a skull of immense size, to which a portion of a so-called horn was attached. I was delighted to see so rare and so precious a thing. I saw at a glance that the skull resembled that of a whale, and had, like the latter, two blowholes on top, which opened into the mouth. I also noticed that what was called a horn, was inserted in the left side of the upper jawbone. Having learned that a similar animal had been captured and carried to Iceland, I wrote at once to the Bishop of Holl, who had been my pupil, and requested him to send me a drawing of the same. He did this promptly, adding that the Icelanders called it narwhal, which means, a whale that feeds on corpses, since whal means a whale, and nar a corpse."

It was, however, a fact, that the Greenland Company purposely defeated all efforts to obtain an animal of this kind; as the fictitious value of a horn of the unicorn was far more profitable to their treasury than the tooth of a narwhal. In 1636, two of their vessels had bought some fine horns from the natives in Davis' Straits, where they had been compelled to winter. Some time afterward one of their agents went to Russia, and offered to sell the Czar Alexis, the father of Peter the Great, two of these precious curiosities, as veritable horns of that unicorn which is mentioned in Holy Writ, and spoken of by Aristotle and Pliny. Alexis admired them very much, and actually offered the enormous sum of six thousand dollars for the finest; but before concluding the bargain, he proposed to consult his physician. This man was learned and experienced enough to examine them properly, and he soon discovered from their structure that they were teeth, and not horns. The Czar

4

dismissed the agent, who returned crestfallen to Copenhagen, and received for his consolation the sneering
question, why he had not first offered two or three hundred ducats to the physician, who would then have seen
as many horns of the unicorn as he could have wished?

III.

PEARLS.

"The chief place among all precious things belongs to the pearl."—PLINY.

A DUSKY fisherman in the far-off seas of India once
found a pearl in an oyster. He had heard of such
costly gems, and sold it to an Arab for a gold coin which
maintained him for a whole year in luxury and idleness.
The Arab exchanged it for powder and shot furnished
him by a Russian merchant on board a trading vessel,
who even yet did not recognize the dirty, dust-covered
little ball as a precious jewel. He brought it home as a
present for his children on the banks of the Neva, where
a brother merchant saw it and bought it for a trifle.
The pearl had at last found one who could appreciate its
priceless value. The great man—for it was a merchant
of the first class, the owner of a great fortune—rejoiced
at the silent fraud by which he had obtained the one
pearl of great price, without selling all and buying it
fairly, and cherished it as the pride of his heart. Visitors

came from all parts of the world to see the wonder. He received them, in his merchant's costume, in a palace plain without but resplendent inside, with all that human art can do to embellish a dwelling, and led them silently through room after room, filled with rare collections and dazzling by the splendor of their ornaments. At last he opened with his own key the carved folding-doors of an inner room, which surprised the visitor by its apparent simplicity. The floor, to be sure, was inlaid with malachite and costly marble, the ceiling carved in rare woods, and the walls hung with silk tapestry; but there was no furniture, no gilding, nothing but a round table of dark Egyptian marble in the centre. Under it stood a strong box of apparently wonderful ingenuity, for even the cautious owner had to go through various readings of alphabets, and to unlock one door after another, before he reached an inner cavity, in which a plain square box of Russia leather was standing alone. With an air akin to reverence, the happy merchant would take the box and press it for a moment to his bosom, then, devoutly crossing himself and murmuring an invocation to some saint, he would draw a tiny gold key, which he wore next to his person, from his bosom, unlock the casket, and hold up to the light that fell from a large grated window above, his precious pet.

It was a glorious sight for the lover of such things. A pearl as large as a small egg, of unsurpassed beauty and marvellous lustre. The sphere was perfect, the play of colors, as he would let it reluctantly roll from his hands over his long white fingers down on the dark table, was only equalled by the flaming opal, and yet there was a soft, subdued light about the lifeless thing which endowed it with an almost irresistible charm. It was not only the pleasure its perfect form and matchless beauty gave to the eye, nor the overwhelming thought of the fact that the little ball was worth any thing an emperor or a millionaire might choose to give for it—there was a magic in its playful ever-changing sheen as it rolled to and fro—a contagion in the rapt fervor with which the grim old merchant watched its every flash and flare, which left few hearts cold as they saw the marvel of St. Petersburg. For such it was, and the Emperor himself, who loved pearls dearly, had in vain offered rank and titles and honors for the priceless gem.

A few years afterwards a conspiracy was discovered, and several great men were arrested. Among the suspected was the merchant. Taking his one great treasure with him he fled to Paris. Jewellers and amateurs, Frenchmen and foreigners, flocked around him, for the fame of his jewel had long since reached France. He refused to show it for a time. At last he appointed a day when his great rival in pearls, the famous Dutch banker, the Duke of Brunswick, and other men well known for their love of precious stones and pearls, were to behold the wonder. He drew forth the golden key, he opened the casket, but his face turned deadly pale,

his eyes started from their sockets, his whole frame be
gan to tremble, and his palsied hand let the casket drop.
The pearl was discolored! A sickly blue color had
spread over it, and dimmed its matchless lustre. His
gem was diseased; in a short time it would turn into a
white powder, and the rich merchant of St. Petersburg,
the owner of the finest pearl known to the world, was a
pauper! The pearl had avenged the poor Indian of the
East, the Arab, and the poor traveller, and administered
silent justice to the wrongful owner.

There is injustice, grievous wrong and fearful cruelty
in the early history of almost all oriental pearls, for, as
Barry Cornwell sings so well—

> Within the midnight of her hair,
>   Half hidden in its deepest deeps,
> A single peerless, priceless pearl
>   (All filmy-eyed) forever sleeps.
> Without the diamond's sparkling eyes,
> The ruby's blushes—there it lies,
> Modest as the tender dawn,
> When her purple veil's withdrawn—
> The flower of gems, a lily cold and pale.
> Yet, what doth all avail?—
> All its beauty, all its grace?—
> All the honors of its place?
> He who plucked it from its bed,
>   In the far blue Indian Ocean,
>   Lieth, without life or motion,
> In his early dwelling—dead!
> All his children, one by one,
> When they look up to the sun,
> Curse the toil by which he drew j
> The treasure from its bed of blue.

For sad is the life and fearful are the dangers through
which the unfortunate pearl-diver passes before his few

years are ended, and he enters into eternal rest. How strange is the providence of God, which places the precious diamond in the hand of the poor Brazilian slave, and grants the precious pearl to the half-starved Indian ! Far out, off the coast of Ceylon and on Bahrein Island, in the Persian Gulf, are the great deposits, from whence come to us most of the gems we value so highly. It is a strange sight to see in the season, in the months of February and March, those desert and barren spots suddenly bloom forth in gorgeous colors, as the sands and coral rocks are covered with tents of richly-dyed materials, and a motley crowd assembles on the forsaken spot. There are divers and merchants, fish-sellers and butchers, boat-caulkers and sail-makers, jewellers and idle talkers, men from Asia and Africa, all talking loudly, jostling each other, eager to become rich by some lucky venture. There are priests also, who levy tribute on the superstitious fishermen, imposing offerings and prescribing holidays, so that the poor fisherman's earnings are half-spent.in advance, and his actual work-days amount to little more than thirty in the season.

When all is prepared, a Hindoo or Parsee blesses the water to drive away the sharks — for a consideration; magicians and sorcerers sell amulets and utter blessings —for a consideration ; and when the boats are ready for a start, there is seen in the chief boat a jolly old cheat, a conjuror and binder of sharks, who waves about his skinny hands and jumps and howls, till the poor fishermen

are as much afraid of his incantations as of the sharks
themselves. They must fast rigidly, while he performs
his wicked rites, nor will he allow them to start, till he
has declared the moment propitious. At last he lifts
up his voice in a hideous way, the divers join in the
chorus, a kind of toddy is made and liberally distributed
among the excited crowd, and the work begins in
earnest.

The boats generally assemble at a late hour of the
night, and when all are together, a signal gun is fired,
whereupon they set sail for the banks, which are not far
from the west side of the Persian Gulf. The purpose is
to reach there before daybreak, so that the divers may
be able to begin the moment the sun rises above the
dark waters. In each boat there are besides the pilot,
ten rowers and ten divers. The latter, perfectly naked,
but with their skin well rubbed with fragrant oil, work
five at a time, leaving the other five to recover and to
recruit in the meanwhile. Before they jump in, they
compress the nostrils tightly with a small piece of horn,
which keeps the water out, stuff their ears with beeswax
for the same purpose, fasten a network bag, which is to
hold the oysters, by a string to their waist, and aid their
own descent by a large stone of red granite, which they
catch hold of with their foot. Then they go quickly
down to the bottom. Here they dart about as swiftly as
they can, picking up with their fingers and with their
toes, which they use with wonderful agility, fill their

bag, and shake the rope that is held above in the boat, in order to be drawn up at once. 🜚

In favorable weather the divers may go down from twelve to fifteen times a day; if the weather is less propitious, they dive at most five times. They remain on an average not over a minute under water; to stay there a minute and a half or two minutes is possible only for a few expert divers, and can only be reached by extraordinary efforts. A few who have endured four or five minutes are spoken of as we speak of the men of genius that adorn a nation's annals; and the greatest of divers is a half-fabulous Indian, who remained full six minutes under water. The exertion is extremely violent, and generally when the poor men return to the surface, blood flows from nose, ears, and eyes. Hence divers are generally unhealthy, and without exception short-lived. They suffer from heart-diseases and sores, and are easily recognized among the mixed population of those regions, by their bloodshot eyes, staggering limbs, and bent backs. These are part of their wages. Sometimes they die suddenly on reaching the surface, as if struck by a shot, and are seen no more. The stories of some of their number being regularly slain, in order to throw their limbs to the sharks for the sake of saving the lives of the others, or of eyeballs starting out of their sockets, and the tympanum of the ear breaking under the pressure of the water, are of course fables; but the pains, perils, and penalties of the poor pearl-divers are, in all

4*

conscience, sad enough to surround the fruit of their labor, the beauteous pearl, with a melancholy interest unknown to other jewels. They have, however, their companions of suffering in higher regions also, for Dryden's words, " He who would search for pearls must dive below," apply to gems more precious even than the costliest of oriental pearls.

The coast of Ceylon, however, is by no means the only place where pearls are found and fished. In the Persian Gulf more than thirty thousand men are employed in three thousand boats, and the produce of their industry constitutes the chief source of income of the Imaum of Muscat. The Red Sea also furnishes a large supply, and these three localities were the sources from which the Romans and the Greeks obtained their pearls. Inferior specimens are found in the Pacific and the West India waters, though certain fisheries on the California coast have occasionally produced very valuable pearls.

It may be assumed, however, that all the mountain-streams of Europe and America furnish a limited number of shellfish, which contain at times valuable pearls. In many small rivers of our mountain regions small pearls have been found, and one of considerable size was a few years ago picked up on the banks of the James River, near Richmond. Certain streams in England have been fished for pearls, in ancient times. Already Pomponius Mela, one of the oldest Latin writers, states that the seas of Britain generated gems and pearls, and Suetonius preserves the tradition that Julius Cæsar was tempted to invade the distant island mainly by the hope that he would enrich himself with its pearls! It seems to be a well-established fact, that the great conqueror brought home from there a breastplate studded with pearls, which he dedicated to Venus Genitrix in her temple at Rome, and on which there was an inscription, stating distinctly that these pearls were British, as Cæsar wished it to be understood that the offering was formed of spoil obtained in Britain. Pliny mentions these pearls as small and ill-colored, but does not doubt their origin.·

Scotland has to this day its successful pearl-fisheries, especially in the river Tay, where they extend from the town of Perth to Loch Tay, and where the mussels are collected by the peasantry before harvest-time, when they enjoy conparative leisure. The pearls, however, are generally small, or, when they are of larger size, rather deformed. It is constantly affirmed by tradition, on the other hand, that the superb pearl in front of the Scottish crown was obtained in the river Ythan.

Pennant tells us that English rivers also were noted for having several kinds of mussels, which produced quantities of pearls, and that there are regular fisheries in many, as in the Esk. In North Wales, the river Conway had, and still has, quite a reputation for its treasures. Camden gives an account of some very valuable pearls found in his time, which he calls as large and as well colored as any we find in England and Ireland, and

adds, that they have been fished for there ever since the
Roman Conquest. Gibson, who translated Camden, says
he knew a Mr. Wynn, who had a valuable collection of
pearls, found in the Conway, among which was a stool-
pearl, of the form and size of a button-mould, and weighing seventeen grains. One of these gems, a Conway
pearl, is to this day preserved in the royal crown of England; it was presented to Catharine, the Queen of
Charles II., by her chamberlain, Sir Richard Wynn, of
Giordir. Even in our day these fisheries are not quite
neglected, but they represent the very prose of the pursuit, as the dangers and difficulties which have to be encountered in the Far East constitute its poetry. As
soon as the tide is out, these simple fishermen go in several boats to the mouth of the river, and there gather
into their sacks as many mussels as they can obtain before the tide returns. These are thrown into huge kettles over a fire, to be opened, and then they are taken
out, one by one, with the hand, and thrown into tubs.
One of the men steps barefooted into these, and stamps
upon them until they are reduced to a pulp. Next they
pour water upon the mass, to separate the fishy substance from the heavier parts, which contain sand, small
pebbles, and the pearls that may have been obtained.
After numerous washings, the sediment is put out to dry,
and the pearls are carefully laid on large wooden plat-
ters, one at a time, with a feather. When a sufficient
quantity is gathered, they are taken to an overseer, who

pays the fishermen a few shillings an ounce for them.
The pearls are generally of a dirty white, and sometimes
blue. What makes this fishery singular is the mystery
which hangs upon the next step in the proceedings. No
one knows what becomes of the pearls. The fishery is a
monopoly, and there is but one person who buys them
up, and as he keeps his counsel most jealously, this has
led to very fanciful surmises. One curious inquirer was
gravely told that all the pearls here found were sent
abroad to be manufactured into seed-pearls, and another
learned that they were exported to India, in order to be
dissolved in the sherbet of nabobs.

Ireland also has its miniature fisheries ; the mussels are
found set up in the sand of the river-beds, with their open-
side turned from the torrent, and contain occasionally
fine pearls. In Bavaria the poor shellfish are treated sci-
entifically : they are put back into certain localities, fed
with a peculiar food, which frivolous critics say is scien-
tifically prepared by the great Liebig, and subjected to a
careful treatment. The success of this curious project
has, however, not yet become public.

The question how the pearls were originally made, led,
in olden times, to many absurd fables, and even the Middle Ages were not free from the wildest theories. Pliny
gravely asserts that the oyster feeds upon the heavenly
dew, and that this produces pearls. Boethius has the
same notion, and speaking of the pearl-mussel in Scottish
rivers, he says : " These mussels, early in the morning,

when the sky is clear and temperate, open their mouth a
little above the water, and most greedily swallow the dew
of heaven ; and after the quantity and measure of dew
which they swallow, they conceive and breed the pearl."
Even Harrison still claims that the pearls are only sought
for in the latter part of August, because a little before
that time " the sweetness of the dew is most convenient
for the kind of fish which doth engender and conceive
them." The common belief in the East is, to this day,
that these precious gems are

> " rain from the sky,
> Which turns into pearls as it falls in the sea ; "

and this is about as true an account of their origin as the
pretty conceit of Robert Herrick :

> Some asked me where the rubies grew ?
> And nothing did I say,
> But with my fingers pointed to
> The lips of Julia.
>
> Some asked how pearls did grow, and where ?
> Then spoke I to my girl,
> To part her lips, and showed them there
> The quarelets of pearl.

Alas for poetry and romance ! The same terrible
science of chemistry which has, with its sledge-hammer
of matter-of-fact, converted the glorious diamond into
vulgar charcoal, has also pronounced the precious pearl
to be nothing but a few layers of membrane and common
carbonate of lime. And yet, here as everywhere in God's
beautiful nature, the poetical element is not wanting, if our
eyes are but opened by wisdom from on high, to see the

daily wonders by which we are surrounded. The pearls, aside from their beauty and their value, are superb illustrations of that beneficent law of Nature, by which injuries are converted into blessings, and Death is changed into Life. The mollusks are all made after the same model, and the common naked snail, as well as the mussel, the cockle, and the oyster, the awkward garden-snail crawling slowly on the moist ground, and the graceful nautilus sailing lightly over the blue waves, the elegant and the rough, the rare and the common, all show the same wisdom and marvellous adaptation of form to their purpose in life. The body is invariably of soft consistence, and enclosed in an elastic skin. From this skin exudes continually a calcareous matter, which resembles common lime. This protects the animal, and serves to form its shell. Where the waves are rough and rocks abound, there this house also is rough, hard, and stony, fit to weather the tempest, and to roll among rocks: where the waters are smooth and only halcyon days to be looked for, there Nature, which never works in vain, provides only paper sides, and an egg-shell boat, such as the little nautilus navigates during his happy life. This same calcareous matter which the animal gives out without pain and without labor, also fills the little house inside with supernatural beauty. It forms that beautiful substance, so smooth and so highly polished, dyed with all the colors of the rainbow, and resplendent with a glorious opalescence, which still charms the eye in spite of its having become so common in all our houses. This is the lining of the shell, the nacre, or in its poetical name, the mother-of-pearl. "The inside of the shell," said old Dampier, the stern sailor with the poet's mind, resembling himself the rugged oyster-shell with the beautiful lining within, "the inside of the shell is more glorious than the pearl itself."

No wonder that with such a beautiful house to live in, the oyster should seem to derive its share of pleasure which is given by the great Maker to all his creatures on earth, from an effort to render its bed always soft and cosy to lie warm, packed in close and comfortably. No wonder that with such a disposition, the animal should become a sybarite, and fret at a crumpled rose-leaf on its ivory couch. Hence, as soon as a foreign substance intrudes by some means or other, the mussel begins to make desperate efforts to remove the irritation. It has no means to resist the intruder; it must do as we have to do where our evils are beyond our powers of resistance; it must submit, and endeavor by the means placed in its power by a beneficent Creator to convert the pain into a pleasure, the grief into a glory. Hence, whatever the cause of irritation may be, the process is invariably the same.

Sometimes a tiny grain of sand or some similar foreign substance slips, in a moment of carelessness, through the opening, and gets between the mantle of the animal and the shell, proving soon a great annoyance. At other times some enemy of the poor, helpless shell-fish goes deliberately to work to destroy it: he fastens himself to the outside, and perforates the shell until he gets within reach of his prey. In such cases, the animal begins immediately to cover the intruding grain with a smooth coat of membrane and a layer of nacre, or to plug the opening in like manner with the same substance, in order to shut out the intruder, and to balk him in his murderous design. These accumulations grow from year to year, and finally form pearls adhering to the inner surface of the shell.

These are, however, not the valuable pearls of commerce, which are always found loose in the interior or imbedded in the soft parts of the animal's substance. This arises from the fact that here the source of irritation has not come from without, but originated in the interior of the shell itself. The cause of this: the animal produces annually a number of eggs, contained each in a tiny capsule of almost microscopic size. As these eggs germinate and become diminutive animals, they are thrown out by the mother, to become mussels in their turn. Every now and then, however, an egg proves abortive, and is not thrown out with the others, but remains behind in the little capsule in which it was originally contained. This capsule, forming part of the animal, and furnished with blood and supplies of every kind by the latter, is gradually covered, like the whole interior of the shell, with nacre, and thus forms the future gem.

This is the way they are made, these wondrous beauties! Well may, therefore, Sir Everard Home exclaim: "If I can prove that this, the richest jewel in a monarch's crown, which cannot be imitated by any art of man, either in beauty of form or brilliancy of lustre, is the abortive egg of an oyster enveloped in its own nacre, who will not be struck with wonder and astonishment?"

All pearls, therefore, have in the centre some small foreign substance, or a tiny cell, which is surprising by its extreme brightness and polish, although but just of sufficient size to hold the original egg. If a pearl be split and then set in a ring with the divided surface outwards, as is often done, a magnifying glass will reveal to us this central cell quite conspicuously, in the form of a round hole, very minute it may be, but well defined, and showing beyond any doubt where the ovum has been deposited. Around this cell an additional coat of nacre is laid evenly and smoothly every year, and thus the beautiful round pearl is gradually built up. Occasionally one may be found that is pear-shaped, and these, when perfect, are considered the most valuable, as they are in great demand for eardrops. This shape arises from the little foot or pedicle to which the egg is attached, being covered with nacre as well as the egg itself.

The great beauty of pearls consists in their perfection of form, and their peculiar lustre, which man has not yet been able to give to artificial pearls, except in rare instances. This lustre arises from two features which characterize these precious jewels of the deep: their transparency and the peculiar structure of their surface. For pearls are transparent, as can easily be ascertained by holding a split pearl to a candle, where, by interposing a colored substance or light, the color will be seen transmitted through the pearl. Now, as the central cell is lined with a highly polished coat of nacre, and the substance of the pearl itself is transparent, the rays of light easily pervade it, and cause that peculiar lustre which characterizes a valuable pearl.

This lustre, however, is heightened into true and superb opalescence by the delicately grooved surface of the pearl, which, Sir David Brewster says, resembles closely the fine texture of the skin at the top of an infant's finger, or the minute corrugations which are often seen on surfaces covered with varnish or oil-paint. In other words, there are, beneath the immediate polish of the pearl, certain tiny wavelets, and dimples, from which the light is reflected in subdued and undulating splendor. From the flat surface of the lining of the shell, the mother-of-pearl, these rays of light diverge in all directions, and hence shine in rainbow colors; in the pearl, on the contrary, on account of its spherical form, the varied hues are all blended into a white, uniform light, which gives to this gem its unrivalled beauty and high value as an ornament.

These lustrous and beautiful spheres are the coveted ornament of all men, and immense prices are paid for

those of perfect form and largest size. Hence man's
cupidity and ingenuity have been at work, from time
immemorial, to imitate Nature's handiwork, and to produce artificial pearls. In the harems of the East, and in
the ball-rooms of Europe, in Chinese homes, and at
American parties, pearls have ere now dazzled the fashion, that never lay in an oyster-bed, as bits of California
rock-crystal have more than once eclipsed the treasures
of Golconda. The result of such labors has rarely been
satisfactory; with the exception of certain French imitations, seen at last year's Exposition, no pearls have
ever yet been produced that could not readily be distin
guished from the genuine product of shell-fish.

It is not a little curious that the nearest cognate substance is bezoar, a concretion of deep olive-green color,
found in the stomach of goats, dogs, cows, and especially camels. The Hindoos generally grind it into yellow paint, but when harder parts are found, they fall
speedily into the hands of jewellers, who polish and
thread them, and then sell them as jewels. Thus it is
from the secretion of a shell-fish, and from the stomach
of lower animals, that man gets the ornaments he most
values for her he loves best, and for him he wishes to
honor most !

Already in the days of the Roman Empire stories were
afloat in the great city, of Arab tribes living near the
sandy shoals of the Red Sea, who practiced the art of
making artificial pearls. They had evidently no inkling

yet of modern ingenuity, for, if we are to believe the Roman writers of the time, these innocent children of the desert went yet to Nature herself for aid in their enterprise, and made the oysters themselves their agents in fabricating artificial pearls. Apollonius tells us how they allured the credulous shell-fish from their cosy bed in the warm waters below to the surface, by pouring oil on the waters, to make them smooth and calm, and seizing them at the moment when they appeared on the surface to imbibe the genial air, thrust a sharp instrument through the gaping valves into the soft body of the animal. Then they threw them into a colander connected with a pan or trough, into which the exuding juices slowly trickled in the form of round pearly drops. The story is, of course, fabulous, but tends to show how familiar the idea of making artificial pearls had already become to the mind of the ancients. The Chinese—that wonderful people, so wise as children, so ignorant in their old age—have likewise for centuries already carried on a well-organized system of manufacturing pearls on the same principle of forced mussel-labor. They claim that this invention was made as early as the thirteenth century, by an individual whose memory they still honor annually by certain ceremonial acts performed in a temple specially dedicated to his name.

The large manufactories of artificial pearls, which now exist near Canton and at Hutchefu, near Ningpo, employ several thousand laborers in this extraordinary business, and produce every year a perfectly enormous quantity of pearls. The process is briefly this: in the months of April and May the full-grown mussels of that year are removed, one by one, from their beds, and have small moulds or forms pushed inside, which are to serve as nuclei for new pearls. A piece of wire or a few metal beads are carefully inserted between the mantle of the animal and the shell, and there these foreign bodies are left embedded in the soft, muscular substance of the living shell-fish, till they become completely incrusted with a thin coating of nacre. A year generally suffices to cover them with a thin but complete coat of mother-of-pearl; but at times they are left much longer undisturbed, in order to obtain a thicker incrustation of greater beauty.

There is in the British Museum a pearl-mussel which has inside the shell a number of little josses made of bell-metal and completely covered and coated with nacre.

The beads so procured have a very handsome appearance and considerable lustre, but they are almost always mis-shapen, following the rough outline of the artificial kernel, and hence they can be sold only for opaque settings, or for embroidery, when the imperfect side is concealed. The principal object of these factories is to produce the small idols with which the Chinese adorn their caps. These are produced by little tin moulds of stereotyped shape, which are inserted into the mollusk, and soon becoming covered with an extremely thin layer of nacre, appear entirely formed of the lustrous substance of which pearls are made. The deception is all the greater as the nacre, though infinitely thin, still forms a complete and unbroken coat of exquisite smoothness, which cannot easily be removed by force, and hence is very durable.

In Europe, it was Linnæus, the great botanist, who first broached the idea of producing genuine pearls by a similar method, and offered, in 1761, to sell the secret to the Swedish Government for a modest sum. The country was, however, too poor to purchase the discovery, which thereupon fell into the hands of a wealthy merchant of Gottenburg. When his heirs a few years later offered the secret, carefully sealed up in the original paper, for sale, it had already become known through the publications of the great savant himself, and all the world was aware that the pearl was the result of an injury inflicted on the body or the shell of a mollusk. Linnæus had, himself, in his collection, several genuine pearls, the forced production of fresh-water pearl-mussels.

The Venetians had long before made pearls in their famous glass-factories. They took hollow glass beads and injected them with various tinted varnishes, into the composition of which certain mercurial preparations entered largely. This manufacture was soon brought to a high degree of perfection, and led to a remarkable evidence of the honesty of the Great Republic: a law was passed by the Senate, toward the end of the fifteenth century, forbidding the sale of these admirable imitations, on the ground that it was fraudulent to make or sell beads which could not be distinguished from genuine oriental pearls! The island of Murano, which was the original seat of this manufacture, has continued until now the principal locality for the production of these artificial or seed-pearls, and their sale is no longer hampered by republican regulations.

The city of Rome boasts of equal success, but achieves it by very different means. Here glass is not so easily obtained, and hence beads of alabaster are carefully turned to a perfect sphere, and then covered with a cement which consists chiefly of finely-ground mother-of-pearl. They do not pretend to compete with genuine pearls, but are an exceedingly pretty ornament, and prove their popularity by never going out of fashion.

The French, whose brass jewels now defy detection, have in the imitation of pearls also proved themselves infinitely superior to all competitors. A few specimens of their artificial productions, exhibited at the Exposition of 1867, could neither in lustre nor in water and color be distinguished from oriental pearls, even when the genuine and the sham were laid side by side. There is but one way by which they may be discovered: this is their specific weight—they are much lighter than the real pearls.

The invention of their composition was, like so many inventions of this kind, due to what is termed an accident. A rosary-maker, in the days of Louis XIV., was walking in the garden of his country-house, near Paris, when his attention was attracted by the silvery lustre on a basin of water. He inquired the cause, and found that a number of bleaks—a small white fish of that region— had been crushed in the water; further examination convinced him that the lustrous sheen was produced by countless scales of the little animals. This suggested to his inventive mind the idea of using the scales for the manufacture of artificial pearls; but at first they decayed too quickly to be of any use. Long reflection led him at last to the happy thought of throwing the scales into a strong alkaline solution, and, lo, the danger was removed! Now there exists large factories where this substance is made. Enormous quantities of the fish, which fortunately abounds in small tributaries of the Seine and the Marne, are caught, and the scales scraped off, well washed in water, and then compressed between folds of fine linen. The fluid which trickles from them is repeatedly filtered until it acquires the necessary degree of purity, and then mixed with some alkaline solution, to prevent the animal-matter that remains from decaying. This is the famous Essence d'Orient, and it takes from seventeen to eighteen thousand fish to make one pound of the pure essence.

At the same time glass-beads are blown with special care so as to produce perfect spheres, and into these the costly essence, mixed with some isinglass, is gently blown by means of a blow-pipe. As if by a magic touch,

5

the glass-bead is instantly changed into a lustrous pearl. They are then steeped in alcohol, dried over a hot plate, filled with wax or cement to give them weight, and finally exposed to various fumes, which constitute the secret of the manufacture.

With all this labor and ingenuity a pearl is produced —a sham. We prefer the workmanship of Nature in the wing-shelled pearl-bearer, the *avicula margaritifera*, a mussel as remarkable for its beauty and eccentricity of shape as for the pearls which it contains. It is now almost exclusively confined to the tropics, though in ancient times it seems to have been found in the Northern Seas also. Its rivals are a mya, which abounds on the shores of almost all seas, and a unio, the British pearl-bearing mussel, found in rivers and small sheets of water. These modest mollusks, unpretending in appearance, but full of precious gems within, produce the pearl which from time immemorial man has valued among the most precious gems; for there are few things so immortal as good taste. Even the inferior pearls have their mysterious value in the eyes of many. The imperfect or discolored pearls are ground up, or dissolved, and used as medicine in Eastern lands. They call the powder Majoon; it is an electuary, and myriads of small seed-pearls are ground to impalpable powder in order to make the costly dose. This is, of course, a mere matter of taste, for the simple lime from the inside of the shell would be in every respect as white and as good, and com-

mon magnesia would have precisely the same effect. But
if some old Emir or rich Bouse is desirous to pay an enormous price for something which he hopes will do his poor
old body good—why should he not be allowed it to do
so ? Have not his betters swallowed everything from
pure gold to toad's brains, from tar-water to the filings
of a murderer's irons ?

The finer pearls, which are not sold on the spot to agents
from abroad, are sent to Europe, and of these the most
valuable find their way, in the course of trade, very
quickly, to London and Paris, where enormous prices are
paid for fine specimens. This mania is, however, by no
means of recent date, for antiquity has its lessons in this
respect also. We all know how Julius Cæsar, when he
was in love with the mother of Marcus Brutus, gave her
a pearl worth nearly a quarter of a million of our money ;
and how Marc Antony drank one, dissolved in vinegar,
which cost nearly four hundred thousand dollars ; whilst
Clodius, the glutton, swallowed one worth forty thousand.
The example of Cleopatra found an imitator even in so-
ber England. Sir Thomas Gresham, not otherwise famous
for acts of folly, still so mistook the meaning of loyalty
that he ground a pearl, which had cost him fifteen thousand pounds, into a cup of wine, in order thus fitly to
drink the health of his great queen ! This plagiarist
again had many rivals in the mad courtiers of Louis XIV.,
who, in their insane extravagance, were wont to pulver-
ize their diamonds, and occasionally used the powder to

dry the ink of letters which they sent to their beloved ones. Is diamond-powder in the hair much worse?

The largest pearl on record is probably one bought by that most romantic of all travellers and dealers in precious gems, Tavernier, at Catifa, in Arabia, where a pearl-fishery existed already in the days of Pliny. It is said—for the pearl is unknown to our day—to have been pear-shaped, perfect in all respects, and nearly three inches long; he obtained from the Shah of Persia the enormous sum of a hundred and eleven thousand pounds for the gem.

Mr. Hope's pearl, which is looked upon as the finest now known, is two inches long and four inches round; it weighs eighteen hundred grains, and, like all such rarities, is of such enormous and uncertain value, that no one would buy it at a market price. The most beautiful collection of pearls belongs, however, to the Dowager Empress of Russia. Her husband was exceedingly fond of her, and as he shared, with other fancies, also that for fine pearls with her, he sought for them all over the world. They had to fulfil two conditions rarely to be met with: they must be perfect spheres, and they must be virgin pearls; for he would buy none that had been worn by others. After twenty-five years' search, he at last succeeded in presenting his Empress with a necklace such as the world has never seen before.

As this admiration for fine pearls has been the common weakness of man in all ages and in all countries, we need not wonder at their playing a prominent part in religious writings. The Talmud has a pretty story, teaching us that those who believed in it, esteemed but one object in nature of higher value than pearls. When Abraham approached Egypt, the book tells us, he locked Sara in a chest that none might behold her dangerous beauty. But when he was come to the place of paying custom, the officer said: " Pay custom! " And he said: "I will pay the custom." They said to him : "Thou carriest clothes." And he said : " I will pay for clothes." Then they said to him : "Thou carriest gold." And he answered them: "I will pay for gold." On this they further said: "Surely thou bearest the fine silk." He replied : " I will pay custom for the finest silk." Then they said : "Surely it must be pearls that thou takest with thee." And he only answered: "I will pay for pearls." Seeing that they could name nothing of value for which the patriarch was not willing to pay custom, they said: " It cannot be but thou open the box and let us see what is within ! " So they opened the box, and the whole land of Egypt was illumined by the lustre of Sara's beauty—far exceeding even that of pearls !

Hence pearls are repeatedly used in Holy Writ also for the most solemn comparisons, and to denote the highest degree of perfection. In the Old Testament wisdom is praised as above pearls, and in the New Testament the kingdom of heaven is compared to a pearl of great price, which, when a merchant had found it, he went and sold

all that he had, and bought it. Even the New Jerusalem was revealed to St. John under the figure of an edi-
fice of twelve doors, each of which was a single pearl.

And this precious gem, fit to adorn an emperor's crown,
and to heighten the beauty of the fairest of maidens—this
pearl of great price, perfect in form and beauteous in
lustre—this jewel of the deep, sought for at the peril of
human life, and paid for with the bread of ten thousands
—it sickens and dies and vanishes in a day. Every now
and then we hear of a noble family, which prided itself
on the possession of magnificent ancestral pearls, panic-
stricken by finding some of their precious gems turning
of a sickly color and crumbling into dust. It is but a
few years since the crown-jeweller of France solemnly
applied to the Academy of Sciences for a remedy against
this disease, caused probably by the decomposition of
the membranes which form part of the pearl, and are
after all liable to decay and corruption, like all animal
matter, by contact with the air. There was no answer
given, but the advice to preserve the precious gems, as
much as possible, from the influences of light and air;
and the Crown of France has since lost some of its most
highly-prized jewels. "Behold, all is vanity and vexa-
tion of spirit!"

IV.

CORALS.

" Unheard by them the roaring of the wind,
The elastic motion of the waves unfelt ;
Still, life is theirs, well suited to themselves."

GLIDING slowly over the blue waters of the Mediter ranean, you often see suddenly beneath you, at no great distance from the surface, a meadow of surpassing beauty. Long green grass, waiving gently to and fro, shine with emerald beauty, speckled with flickering lights; and all over the little prairie are scattered flowers in brilliant colors. The restlessly heaving water increases the splendor of the scene ; and dazzling hues of green, orange, and deep red, shine upward through the transparent waves. But the oar splashes, and in an instant all the beauty of coloring has vanished, and the whole region is clad in a dull dingy gray. You become aware that you are in the midst of a colony of animals, so small that the naked eye can hardly discern them, and yet so powerful, by the strength of their united forces, that they have built whole islands in distant oceans, and

raised lofty mountain ranges in the very heart of
Europe. But they are most sensitive little beings, and
the slightest touch of a foreign body, a single ray of the
sun, or an angry splash of a headlong wave, makes them
shrink back into their narrow home.

They are altogether a strange, mysterious race, these
Maidens of the Sea, as the ancient Greeks used to call
them. Their beauty of form and color, their marvellous
economy, their gigantic edifices, all had early attracted
the attention of the curious, and given rise to fantastic
fables and amusing errors. They were well known to
the chosen people, for, singing of the grandeur of Tyre,
the prophet states that "Syria was thy merchant by
reason of the multitude of the waves of thy making:
they occupied in thy fairs with emeralds, purple, *coral*,
and agate;" and ancient Job even mentions coral
among the most precious things, and yet was not fit to be
mentioned in comparison with wisdom—thus proving the
high value which already in those early days was attached to the red corals. We learn, from other sources
among profane writers, that priests wore them as amulets, and physicians prescribed them in many diseases as
useful remedies; whilst Pliny enters into a more detailed
account of the manner in which they were used for purposes of ornamentation, how weapons were adorned with
them, and costly vessels derived additional value from a
few deep-red branches of the Flowers of the Sea.

For flowers they were held to be from time immemori-

al, and for centuries even of our Christian era. These
bright-colored, delicate forms, which, taken out of their
element, changed miraculously in an instant into dingy
brown stones, were believed to be real water plants, which
the contact with the air turned at once into stone. Nor
is this belief extinct among men: the dwellers on the
coast of Southern Italy still swear to it, and laugh in their
beard when the foreign savant speaks of them as life-
endowed animals. It seems now astonishing how men
could quarrel so long and so pertinaciously over the apparently simple question, whether corals belonged to the
vegetable or the animal kingdom. More fortunate in
this respect than many other organic forms, whose social
status is not yet recognized, corals were already, in the
beginning of the last century, raised to the dignity of
animals. This was not achieved, however, without much
trouble and much ludicrous blundering. It was a young
physician from Marseilles, called Peyssonel, whom the
French Academy had sent to the coasts of Barbary, for
the purpose of studying salt-water plants, who first discovered their true nature, and observed how they ex-
panded and contracted at will and moved their arms with
a purpose. He communicated his discovery to the great
Réaumur; but the illustrious naturalist was still so firmly
bound by precedent and scholastic method that he refused
to endorse the bold doctor's statement, withholding, however, with equal courtesy and discretion, his correspond-
ent's name; for what is now praised as a noble progress

in science, appeared to him a rash statement, likely to injure the growing reputation of his young friend. It was only after an interval of twenty years, when Trembley had published his beautiful discovery of sweet-water polypi, and Jussieu, the master of botany, had given to corals their papers of dismissal from his kingdom, that Réaumur made the *amende honorable*, and acknowledged both the correctness and the great value of Peyssonel's discovery. But where was the victim of his previous reluctance to appreciate his merit? He had gone, in disgust and despair, to the West Indies, and there he had disappeared from the sight of men, so that to this day we know neither the time nor the place of his death.

Since then we have learnt much, but by no means all yet, about the birth, the life, and the end of corals. In the hot summer-months, when the waters are bringing forth, as in the days of the creation, the moving creature that has life, millions of diminutive, jelly-like spawn are thrown out by the parent animal. For awhile they enjoy their freedom, and seem to luxuriate in the exercise of their powers of locomotion, which they are never hereafter to recover; but soon they become weary, and settle down upon some firm, stationary body. At once they begin to change their form; they become star-like, the mouth being surrounded by tentacles, very much as the centre of a flower is surrounded by its leaves. After some time, each one of these ray-like parts pushes out extensions, which in their turn assume the shape of tiny stars, and establish their own existence by means of an independent mouth. In the meanwhile lime has been deposited at the base of the little animal, by its own unceasing activity, and forms a close-fitting foot, which adheres firmly to the rock. Upon this slender foundation arises another layer, and thus, by incessant labor, story upon story, until at last a tree has grown up with branches spreading in all directions. But where the plants of the upper world bear leaves and flowers, there buds forth here, from the hard stone, a living, sensitive animal, moving at will, and clad in the gay form and bright colors of a flower.

This flower is the animal itself, seen only in its native element, and unfit for air and light. What we call coral is its house, outside of which it prefers to live rather than within. How they build their dwelling, human eye has never seen. We only know that the tiny animals, by some mysterious power given them by the same great Master on high who has given us a body after his image, and immortal souls, absorb without ceasing the almost imperceptible particles of lime which are contained in all salt-water, and deposit them, one by one, in the interior. This is done now more, now less actively; and the denser the deposit is, the more valuable the coral. Gradually this substance hardens and thickens, until in the precious coral, the *Isis Nobilis* of science, a large tree is formed, which often reaches the size of a man's waist. It is perfectly solid and compact, and

adorned on the surface with delicate, parallel lines.
Thus on the tree-shaped limestone grows the life-
endowed body of the polypus ; it moves, it feeds, it produces others, and then is turned again into stone, burying itself in its own rocky house, whilst on its grave
new generations build unceasingly new abodes.

This is the so-called Blood Coral of the common people, the favorite of antiquity, and the fashion of our day
—next to the pearl, the most precious jewel of the
deep.

It is not easy to obtain a piece of living coral, for the
purpose of studying its wondrous structure and admiring its exceeding beauty.   The great depth at which the
mysterious little animals dwell in the ocean secures them
against the mere amateur fisherman ; and the professional
coral-fisher, the son of superstitious races in Southern Italy,
is extremely reluctant to admit outsiders into the secrets
of his trade.   If you ask him to bring home for you a few
valueless pieces, he is afraid of witchcraft, and the vessel
you have given him for the purpose is filled with every
animal from the deep but corals.   If you follow him in
your own boat, as he sails out for his day's work, he is
more seriously frightened still, and takes to the open sea
—preferring to lose rather a whole day's labor and profit
than to betray his favorite fishing-ground.   He cannot
comprehend why you should be willing to pay him well
for what has no value in his eyes; and, like the Arab
who suspects every travelling Frank of seeking after con-

cealed treasures, the poor Neapolitan fancies you possess a charm by which you can change his shells and sponges into precious pearls and corals. Even after you have succeeded in persuading him that you are no sorcerer, and never studied in the school of that great magician, Virgil, he fears you may betray the few, simple mysteries of his trade, or the locality, from which he derives his support. It requires much time, much money, and especially much patience, to convince him of your innocence, and, even when all these obstacles are removed, he still pertinaciously adheres to his hereditary superstition, that it is of no use to try catching corals alive, as they are sure to die of fright as soon as they behold the light of day. Hence it was by an accident only that I was fortunate enough once to see how corals are fished, and to examine them closely, when fresh caught.

It was a Sunday, and we were sauntering up to the tall olive-trees of St. Hospice, near Nice, in order to enjoy there our self-caught meal of lobsters and cuttle-fish, when we suddenly caught sight of an odd-looking craft lying far out in the beautiful bay of Villafranca. The sails hung carelessly about, and the bowsprit stood bold upright, being crowned at the top with a couple of saints carved in wood, while below two huge eyes were painted on the waist of the vessel.

"It is a coraline," said one of our party, an Abbé, familiar with all the features of the country; "poor people, who will stay here many weeks, catch nothing,

spend all they have, and finally sell or pawn their boat
to enable them to return home."

"They have come for the great coral-tree," said our
boatman, who was carrying the hampers. "You know
the one that grows down in the dark grotto near Mount
St. Alban. There is no year that some Neapolitans or
Sicilians do not come up here in search of the treasure,
but no one has ever yet found it."

"Can you imagine," asked the Abbé, "that these
people really believe in an immense tree of coral, which
grows a hundred fathoms below the surface of the sea in
a grotto, large, like an ancient oak-tree, and stretching
out its gigantic branches in all directions, but drawing
them in instantly, like a cuttle-fish, when a net comes
near? That is the story here, and these poor fishermen
believe in it as firmly as in their Holy Virgin, and laugh
us to scorn when we attempt to reason with them, and
prove to them the impossibility of such a thing."

We made up our mind, on the next day, to go on
board the odd-looking boat, and to see what could be
learnt from the crew. Fortunately one of our party
was a Neapolitan, well known to all the fishermen on the
Chiaga, and speaking their curious dialect. By a number of masonic signs he made himself known; and the
air of mistrust and repugnance with which he had at first
been received, gave way to a less suspicious manner.
The padrone, or master of the vessel, was an oldish man,
with a deeply-furrowed face, and a hard expression about

the mouth, which did not promise a very mild government on board. They are a strange class of men, these padroni of coral-boats, hundreds of whom come annually from Naples and Sicily, from Genoa and Sardinia, and sail, with the exception of a few adventurers bound from the coast of France, along the coast of Algiers in search of precious treasures. How on earth they manage to sail, far out of sight of land, without telescope or compass, and there on the broad ocean to find, year after year, the precise place where, far down in the deep, there lie vast masses of rock, which contain in cleft and crevice the desired coral branches, is more than ordinary seamanship can explain. Three things only they need to aid them in these venturesome journeys, which recall to us forcibly that first great search after the Golden Fleece: money in large sums for the outfit of their coralines, a good stock of falsehoods to screen their real purposes, and an invincible silence to oppose to all direct questions. They have a saying among' themselves, that purse, and falsehood, and silence, must all three be as deep as the sea in which they mean to fish.

Our padrone owned himself his little vessel, which did not measure over five tons; his son, a clever, restless little scamp, served as ship-boy; and three sailors sufficed to handle the nets and to work the boat. They had come across the Mediterranean from Torre del Greco, near Naples, in search of the fabled giant-tree, which his favorite saint had shown the padrone in a dream. There was no log and no compass on board, and all their provisions consisted of the never-failing galetta, a white ship-biscuit, and some water; for there is no cooking on board these coralines. The padrone was proud of having a few onions and some dried fish in a locker, the key to which never left the lucky owner's pocket.

We found that the fishing was done with a large net, fastened by a stout rope to the stern of the vessel. At the end of this rope hung first an iron cross, consisting of two hollow tubes laid cross-wise, through which strong ash poles had been thrust, and to this were fastened a number of old sardine-nets, no longer fit for their first purpose, and countless ends and bits of wide-meshed pieces of rope, as thick as a finger—the whole apparatus a mass of rags and rotten net-work. But the more such wretched-looking pieces of net-work the padrone can fasten to his iron cross, the better are his chances. When the sea is perfectly quiet, he lets them sink down to a depth of sixty or even a hundred fathoms, where they slowly spread and unfold themselves over a vast extent. Then he hoists his lateen-sail and slowly drifts before the wind, or, in a calm, sets his men to work at the huge oars of the vessel. If not so engaged, they stand watching at the sheets, the oars, and the tiny capstan, to obey instantly his orders. His one great purpose is to wrap as large a number of his fluttering pieces of net-work as he can around the branches of coral below, to tear them by main force from the parent stems, and to wind them up, together with the fragments of rock to which they are attached.

The padrone seeks by the aid of his mysterious science a favorable spot where corals grow, and his delicate and experienced touch feels instantly, by the gentle stretching of the rope, when the net has caught hold of coral branches. The little vessel, no longer obedient to sail or rudder, is held in check by the stout rope, and hence jumps forward and backward as the net seizes and lets go again far down at the bottom of the sea. The work is hard, and the perspiration is running down the neck of the poor sailors. At times the nets are caught between rocks, and the boat must tack and vere in all directions to try to loosen them ; at other times the padrone makes desperate efforts to creep in between overhanging rocks, into a narrow cleft, for there, in eternal shade and almost inaccessible recesses, they believe they find the largest and most valuable coral branches. Thus they try and drift along, they work and toil and draw up perhaps twenty times a day, and each time it is a mere lottery. And this is the very charm which this kind of fishing has for the poor children of the South ; they hope and hope on, and, sick or well, old or young, not one of them would, when the season comes, willingly give up his chance of finding some precious tree that is to make him rich for life.

At last the padrone thinks he has a net full. The

sailors, whistling a tune through their teeth, man the capstan and work with their hearts beating—a jerk, and the net is loosened and comes up slowly, slowly. All eyes are eagerly bent upon the place where it will appear on the surface; at last it shines with a white gleam, far down still. If the pieces of net-work appear wide spread, the evil omen is greeted with muttered curses: "*Dio grazia! Maledetto!*" If they hang straight down, heavy-laden, the deepest anxiety is seen in all features, and the excitement becomes intense. Now it shines reddish! "*Santissima!*" exclaims the master, and the men work with renewed energy. At last it is alongside. It is heaved on board with great care, and now comes the task of picking out the precious treasure from the meshes of the net-work, and to loosen them from the fragments of stone on which they were growing.

With these stones a thousand odd and outlandish citizens of the deep are curiously intermingled. Here hang worthless horn-corals, and among them the Black Hand of the sailors, which they love dearly in spite of its uselessness, because it is an unfailing sign of the presence of genuine coral. There come up sepia-fishes with staring eyes, long waving arms, deformed bodies, biting beaks, and mighty suckers, abounding in weird and ghost-like shapes. Between these frightful forms wave sea-weeds with broad, green, and purple fronds; while little tufted bunches of red and white and violet and yellow lie marvellously close to feathers, crusted all over by the salt sea-wave. Elfish faces, with huge staring eyes, peep at you from every side, and seem to threaten you with wild, unearthly horrors if you dare touch them. A fulness of strange things, unseen and unsuspected by the dweller on firm land, comes thus forth from the hand of Nature, in her great workshop of the unfathomable, fertile sea. But they are all pitched overboard; only, the men are sure first to open the shell-fish and to swallow the contents with truly marvellous dexterity, before the shells are allowed to return to their dark homes below. The branches of coral are carefully picked out down to the smallest fragment, and great is the joy of the lucky finder if he discover a piece naturally bent in the shape of a little horn, for it is an amulet, a sure protection against the dire effects of the Evil Eye. The whole is thrown into a large chest, the key to which the padrone wears hanging around his neck along with a tiny bag of holy relics; and if there should be a peculiarly thick branch among them, he places that in some mysterious hidden corner, for it is very valuable, as the price of coral increases almost in geometrical proportion with its size.

When all that has been fished up is saved, the boat returns to the harbor and delivers the result of the day's labor to an agent, who carefully and judiciously assorts the pieces according to size and color, and sends them at once to Naples, Leghorn, or Genoa, where they are quickly worked up into every kind of ornament.

But woe to the poor sailors if the net should come up

empty, or, worse still, if it should catch at some projecting point of rock, and refuse to come up altogether! It
is they alone who are blamed; it is they who have, by
their idleness or their wickedness forfeited the favor of
saint and madonna, and who must now labor and toil
until exhausted nature refuses to sustain them any
longer.

The only way to examine the living animal is to seize
the little fragment of rock, or the shell to which the mysterious creature is fastened, at the very moment that it
appears near the surface, and to dip it, if possible, without exposing it to the air, immediately into a vessel with
salt-water, which you hold ready for the purpose. At
first there is nothing to be seen but a vague indistinct
mass of grayish substance. You suspend the animal and
its tiny abode by a string in the middle of the glass globe,
and carry it to a dark place; for the coral will not display its beautiful form and color in the gleaming light of
the day. It takes hours often before the obstinate little
creature condescends to give a sign of life. At last you
fancy that the club-shaped extremity of the dingy red
substance begins to wrinkle up into little rings here and
there. You take up your magnifying glass, and you see
with joy and satisfaction that the eight star-shaped in-
dentations, which mark the diminutive wart-like rising,
assume a white tinge, contrasting pleasantly with the red
at their base, which grows every moment to a more
lively hue. The lines widen and show an opening be-

tween two bright-colored lips; a vague, undefined substance rises slowly, like a transparent globule, but soon
it grows and swells, and at last it stretches out eight
bright, leaf-like arms, edged all around with delicate
fringes. Now the whole resembles strikingly a beautiful flower of eight leaves, not unlike a gentian or a cam-
panula, and you acknowledge at once how pardonable
was the error of those who, for generations, insisted upon
believing the strange animals to be nothing more than
submarine flowers, endowed with the power of motion.

The colors are brilliant beyond all the art of man can
produce. In the true coral a resplendent, almost dazzling red surrounds the base of the bell-shaped body of
the animal, whilst the latter itself, and the arms, appear
as if carved out of transparent crystal. And as soon
as one of the diminutive creatures, bolder than the rest
—or more hungry—has set the example, the others follow in rapid succession, and soon the whole little branch
is covered with living flowers, crowding each other so
closely that it seems as if they would prevent one
another from unfolding. Flowers, however, are still and
motionless; here all is full of life and activity. They
move in slow, measured ways, it is true, but with what
variety! Now the beautiful carolla looks like a half-
opened bell, with its delicate white leaflets rising out of
a deep-red crown; now again it resembles an urn with
faintly-drawn outlines of classic purity, and then it
changes into the shape of a wheel with eight spokes.

As you are still gazing and marvelling at all this exuberance of colors and beautiful forms, which the bounty of the Creator has bestowed upon the dweller in the deep, far below the warm light of the sun and the admiring eye of man, you touch the vessel that holds these wonders, and in an instant the scene is changed. Quicker than the eye can follow, the fringes that adorn the arms disappear, the arms fold themselves up and draw in toward the centre, where the mouth was but just now standing open, ready to receive its invisible food, the beautiful bell is shut up, and the bright-red lips close once more, so that there is nothing left again but the insignificant little branch of dingy color. They are evidently most sensitive little creatures, these strange little animals; like true children of the dark deep, they can bear neither heat nor light, nor the slightest touch of a foreign body, and although they close in an instant, they dare not unfold their beauty again for hours.

In spite of this delicate sensitiveness, nothing looks apparently more simple than the structure of these polypi. Each one is firmly seated in the red, leathery substance, in which his tiny cell is hollowed out. By means of his moveable arms and their cilia, he creates a little whirlpool before his mouth, and seizes the infusoria that serve him as food, together with the particles of lime which he needs for his house. Whatever thus enters passes down into the common receptacle, where it is digested; for the coral polypus is not only a sociable animal, fond of living in large numbers together, but he is a perfect socialist and communist. It is only by the common labor of thousands and tens of thousands of these diminutive beings that the coral branch can be formed, which is to become in the hands of man a jewel of priceless value. This result, moreover, can only be obtained by the readiness with which each individual surrenders the fruit of his labor for the benefit of the whole community. Each polypus, ever busy with its eight agile arms, works night and day, catching as many tiny things as he can seize in the water. He takes the first taste, as of right, throws out all that is unfit for his purposes, and then sends the surplus down into the common stock, from whence it is afterwards distributed equally, through countless channels, into every part of the living tree. The common substance, which serves as highroad for what comes and goes, and in which dwells, so to say, the life of the community, is the thick, red bark which covers the cells of the polypi—not a skin to cover and warm the little animals, but the very mother and nurse of the whole stock, the bond that holds them all together, and the place where the inner, solid kernel is made, which supports the whole tree. Through a thousand little openings and wide-meshed net-works passes the nutritious juice of milky whiteness, which oozes out if the covering be cut, and which the fishermen hence call coral-milk. It is propelled onward and upward by microscopic cilia, similar to those in the inner vessels of the human body.

Thus here also the astounding wisdom of God is beautifully displayed, and the almost unknown body of the
stone-animal is as fearfully and wonderfully made as
that of man himself.

Far away from the Mediterranean, in the vast waters
of the Pacific Ocean, and amid the South Sea Islands, a
kindred race of the true coral, the Madrepores, have been
at work for countless generations. They are the humbler
brethren, unadorned with beauty and unable to furnish
man with costly jewels. But as everywhere in Nature
the humbler is the more useful, and the smaller the more
powerful, so here also. These corals have raised, by in-
defatigable labor, colossal structures, by the side of which
the walls of ancient Babylon, the Chinese wall, and the
Pyramids of Egypt, dwindle into dwarfish proportions.
Amid the most violent storms, and in the most agitated
seas, where wind and waves would easily destroy the
grandest works devised by the skill of man, they erect
their marvellous edifices—architects so feeble and insignificant, that, when they are drawn out of their elements,
they vanish, and can hardly be perceived.

Their works are works of beauty. Like enchanted islands, these circular coral-reefs bask in the brightest light
of the tropics. A bright green ring encloses a quiet inland lake; the ground is white, and, the water being
shallow, it shines brilliantly in the gorgeous floods of
light that fall upon it, whilst, outside, the dark black billows of the angry sea approach in long lines of breakers,

tossing their foaming white crests incessantly against the impregnable ramparts. Above there is a clear blue heaven, and, all around, the dark ocean and the hazy air blend harmoniously into each other. The contrast is beautiful beyond all similar scenes: within all is peace, and soft, mirror-like beauty; without, all is strife and eternal warfare. But the battle is here emphatically not to the strong. The small and lowly polypi, by whose united labor and strength these colossal walls have been raised to say to the ocean, " so far and no farther thou shalt go ! " defy the mighty waves. Year after year, generation after generation, the tempest beats upon their fragile homes, and the mountain-like waves of the ocean come thundering on, like armies of giants, to rush upon the slender reef. But ever and ever the attack is re-pulsed, and the minute animals work quietly, silently, with modest industry and untiring energy, at their heaven-appointed task; and the living force, though so small, triumphs victoriously over the blind, mechanical force of the furious waves.

Their great works either stretch out far into the ocean like huge barriers in continuation of the natural coast, or they form gigantic rings of rock, upon which plants soon spring up, soil is formed, and at last a habitation is pre-pared for man himself. The little polypi find themselves there in strange company. First there are the only ene-mies which they are as yet known to have. Outside the reef, as well as inside of the lagoon, but always within

6

reach of the coral rocks, large shoals of small fishes are
found, which actually feed upon the pulpy, leathery substance of the polypi, and secrete the indigestible material it contains, thus producing a kind of calcareous pulp
which soon changes into fertile soil and serves as an excellent ground for palm-trees and other plants. But as
the eater is always eaten in all nature, so here also the
avenger is at hand. Hungry dogs are waiting patiently
on the reefs and shallows, and as soon as the voracious
fish rises to the surface, where alone he finds the tender
polypus, they pounce upon their prey and swallow it
eagerly. On shore, where the graceful palm-trees flourish,
a feast is provided for another class of hungry claim-
ants. With a heavy thump a cocoanut falls upon the
hard ground, shaken down, before it is perfectly hardened, by a sudden gust; at once land-crabs are seen hur-
rying up at the sound of the simple dinner-bell, and one
of them seizes it, bores with its long, sharp claws into
one of the eyes, where the shell is softest, and sucks with
delight the sweet, milky juice. In light, fragile boats,
daring Malays come from far and near and dive into the
thick grove of coral trees, where they are sure at all
seasons to find a valuable variety of turtle, which often
reaches the enormous weight of a hundred and twenty
pounds. As soon as they perceive one of these monsters, they chase it and try to drive it into shallow water,
or at least, by constant, skilful hunting to and fro, to
exhaust it, so that they can approach quite near. As

soon as this has been accomplished, an active, agile man jumps upon the back of the turtle, takes hold of the slender neck, and, riding boldly on his strange mount, profits by the first firm footing he can obtain to use it as a lever and to turn the animal on its back, when it falls an easy prey to its captors.

Thus plants and animals, and man himself, all owe their food and their life to the tiny coral. But, small as it is, its power of production is so enormous that these diminutive polypi could long since have filled up the basins of the great ocean, and covered the continents of the earth, if their existence were not limited, by an all-wise Providence, within certain local boundaries and fixed conditions of temperature. For, with the exception of a few rare varieties, corals cannot live where they are not permanently covered with water, or at least continually bathed by breakers. Nor can they exist below a depth of about two hundred feet, partly because the enormous pressure of the weight of the water above them would be fatal to all kind of life, and partly because they require a higher temperature than that which prevails at so great a depth. The bright colors in which most of these polypi are clad prove, moreover, that light, the gay painter of nature, is indispensable to their existence, and this element they can only enjoy in the regions nearer the surface. But, above all, being citizens of the animal kingdom, they need, like all animals, oxygen to support their life, and this food is not attainable where the air cannot impart it to the water directly by contact with the surface, or send it, by the agitation of the waves, down to a certain limited depth. Thus we meet here also with those great immutable laws, by which the Creator of all things has assigned to each one of his creatures its abode upon earth, and bestowed upon it powers of grateful enjoyment. Great and wonderful are His works, teaching us everywhere, on land and at sea, on the mountain-top and far down in the depth of the ocean, not merely to glance at the surface, but to look down into the deep, where the costliest jewels are hid in the dark abyss; nor merely to glance at the clouds and the sky, but to lift up our eyes to the heavens, where there are wonders yet awaiting us that " eye hath not seen, nor ear heard, neither have entered into the heart of man."

V.

THE KNIGHT IN ARMOR.

" And, like a lobster boiled, the morn
*From black to red began to turn."—*BUTLER.

THERE are few subjects of deeper interest to us, and
few that present greater difficulties, than the precise
relation in which man stands to that creation which was
assigned him by the Creator as his domain, and which
" groaneth and travaileth in pain together " with him in
a common cause and a common hope of redemption.
When little was known about the lower creatures, judgment was easy ; but as information increased and wonder
after wonder was reported concerning the rare powers
and strange sagacity of many animals, increased interest
led to deeper research, and finally skill, instinct, and even
reason, were assigned to certain families. The most recent theory, the favorite of German savants, is to divide
all created beings, endowed with animal life, into classes
independent of their outward form and nature, according
to their inner, psychical life.

There can be no doubt that the universe was created

not for man alone, but for that "whole creation" with which he finds himself by divine will indissolubly connected. But this great world is not equally open to all, and its unnumbered inhabitants are most variously endowed with means of contact. Some have numerous organs by which they place themselves in communication with the world around them; others appear almost hermits, unwilling to enter into relations with others. But it is evident, even to the superficial observer, that no outward organs are ever given without an inner sense to render the perceptions of the former useful. How the connection is established between the outer sense and the inner reflex is a mystery of mysteries in the most perfect of beings, in man himself—how much more in the animal of whose soul we know so little! This secret belongs to the things which were too wonderful for the wisest of wise men, to the "way of an eagle in the air, the way of a serpent upon a rock, the way of a ship in the midst of the sea." It is the secret which has led the materialist, in his despair, to deny the existence of a soul altogether, which makes the trifler content with the term Instinct, that says nothing, and which leads the faithful believer to the one great source from which alone come wisdom and knowledge.

This much, however, has been ascertained, that each organism is fearfully and wonderfully adapted to the life of the owner, and yet also in complete harmony with that part of the outer world with which it is placed in rapport.

Organism, soul, and world, constitute thus an indissolu-
ble trinity, and nothing excites the marvel of the student
of natural history so much as the infinite wisdom displayed in this union. The smallest of infusoria is, considered in this light, as perfect as the eagle that soars in
the clouds and gazes undazzled into the face of the sun.
Every class is equally perfect in itself; no instinct ever
grows, no powers of discrimination are developed; man,
animal, and plant are, as far as their relations to nature
are concerned, made, once for all, after a perfect pattern.
Poeta nascitur, non fit, is here as true as in poetry. But
with every class new powers are seen to be given, new
bonds established between their inner life and outer
nature. The worm in our intestines discerns nothing but
food or starvation; the butterfly knows colors, the eagle
distinguishes men and animals, and man himself knows
the past and the future as well as the present.

According to this principle, the lowest animals are
those which know least and distinguish least; others,
more favored, have more numerous points of contact
with the world, and the highest orders distinguish, comparatively speaking, all that surrounds them. As there
are even among men whole classes who cannot hear certain very high notes, or see some of the colors familiar
to others; as some minds soar freely and intelligently
into the highest regions of thought, while others are
unable to rise above the common things of the earth, so
there are, no doubt, still higher beings around us, who

discern much that escapes our duller senses and our inferior mental powers.

But where is the line to be drawn, and what symptom
is to be chosen as the standard by which to measure the
rank of each class of beings in the great realm of nature? Here it has occurred to some of our savants, that,
with all the marvellous diversity of form and endow-
ment, no creature yet comes forth fully made and developed, when it first enters the world of life. All, on
the contrary, must begin at the lowest end, and painfully,
slowly, make their way upward. Man himself, made
after the image of the Most High, and but a little lower
than the angels, begins his career as an almost invisible
atom, a shapeless egg. Nor is he alone in this. Every
living being commences the earthly existence as a germ
or an egg, and is then asleep! Here was found the
common feature of all creation—sleep. Sleep is the first
condition, from which dates all life. It is not death;
for there is in sleep already some power of discernment,
not from reasoning but from sensation. Men and animals alike turn towards heat when they feel cold, and
towards the cold air when they are hot, although in deep
sleep. A person, fast asleep, carries instinctively, as we
call it, the hand to the point where he is unduly touched.
Others, suffering from thirst, will go to drink and yet
never awake nor remember afterwards what they have
been doing. Man, being the highest, shows also in pro-
found sleep the finest distinctions. When two persons

suffer or do the same thing in their sleep, they yet act
not alike but show even then the difference in temper
and character. Sleep, however, is almost passive and
the child of darkness. It has nothing in common with
light, but shares with its brother Death the love of
night. There is but a short step from sleep to death—
do we not all fall finally asleep, when we leave this
earth? Hence sleep has no active powers; vegetation,
reproduction, and nutrition, alone continue as long as it
holds the living being captive. Now there is a class of
animals who sleep during all their life; to awake is to
die for them. These are the lowest in the scale of beings; they merely exist, feed, and reproduce themselves, but their soul is dormant. Such are intestine
worms, beautifully made in their adaptation to their
peculiar mode of life, but doomed to live in eternal darkness and seclusion. Bring them to the light, let the free
air of heaven blow upon them, and they die at the instant.

The dreamer is no longer fast asleep. His fancy is excited and certain powers of his inner being are actively
engaged in forming images of the real world without.
But not the imagination alone is at work; there must be
more, since it is possible to make mistakes in dreams.
The impossible is strangely mixed up with the possible.
The dreaming hound runs madly after the fox without
advancing; the horse becomes excited in dreams, and
snorts as if it sniffed the battle from afar. Men perform

mental feats in their dreams, of which they would be incapable when awake, and even the Lord spake of old in dreams to his favored children on earth. Dreams, however, are as yet nearer to sleep than to waking life; they prefer, if not absolute darkness, the more genial twilight. They represent the first budding forth of the tree of knowledge. Hence children begin to dream in their earliest infancy—some say they dream in their mother's womb. But their activity is as yet dim and instinctive; they see the outer world only as through a bright-colored veil, and respond only faintly to impressions from without. As a pressure upon certain parts of the body produces invariably the same dream, so appeals to the ear are understood by the dreamer; he replies to them or he embodies them in his dreams. Hence his intelligence is at work; he is by turns a king and a beggar; and the dreaming animal, like the cat under the influence of valerian, evidently uses its highest powers. Here also the correspondence with the state in which large classes of animals perform all their functions is striking. They live in a dream, and become only vaguely conscious of their relations to the outer world.

Higher than the dreamer stands the somnambulist—not the man with the diseased mind and disordered functions, who is exhibited to a gaping crowd by a charlatan—but the man endowed with the mysterious gift of performing, while apparently asleep, actions which require the wisest judgment and the soundest reflection. His condition is no longer merely passive; he becomes active, and the outer world is evidently quite apparent to him, though he walk with closed eyes. He writes at his desk and arranges his books, or he milks his cows and carries the pail indoors; he walks on the frail gutter hanging from the eaves of a house, or along the brink of a precipice, where no waking man would dare to venture. There are somnambulists who even speak through half-closed lips, and answer rationally to well-considered questions. The link, which here connects the outer world with the inner consciousness, is as yet altogether beyond the ken of man, since even the senses are locked in apparent sleep, and yet impressions are made on the mind and the heart.

There is a large class of animals, those most intimately in contact with men, who are genuine somnambulists. The bee builds her cell, the ant works at her house, the bird fashions his nest, by some such power. It is this manner of life which, no longer a sleep nor a dream, but neither as yet a full waking activity, makes them cling to man, and attaches the horse and the dog to their owner. It is this mysterious power which we most commonly call instinct, by which the brute knows its owner, discovers untaught the remedies it needs, when sick, and which finally culminates in the second sight ascribed to dogs and to horses.

The highest state of life which we know is the waking life, in the full blaze of noon. Man himself, capable of becoming a Plato or a Newton, and striving to follow the example of Him who became the Light of the world, still sleeps and may be a somnambulist. But he does so only because his body requires it, and the inspiration of the poet, the energy of the statesman, or even the determined will of the lowest amongst us, gives us strength to remain awake for days and nights together. Man alone forms this class of created beings, and yet he dreams of still higher modes of life, given to invisible fellow-creatures of his, whose existence he merely presumes, but whose influence he is willing to acknowledge under the name of evil spirits or guardian angels.

Whatever the merit of such a division may be, there can be no doubt that it leads to a more careful investigation of what may not inaptly be called the inner life of animals, and as we have seen that this is invariably represented outwardly by corresponding organs, it lends a new interest to the study of certain individual characters among the different classes. As we have on a previous occasion endeavored to sketch the life of one of the best-endowed of higher animals, we propose here to give an outline of one of the lowest, in whose remarkable form and strange character we yet find new evidence of the Supreme wisdom that made him also useful to others and endowed him with sources of happiness and simple enjoyment.

He dwells far down in dim twilight, among sorrowful brethren, whose homes are the dark earth and the great deep. There is no beauty of color in the dismal waters in which he spends his checkered life; there is no comeliness of shape to be seen in his friends and neighbors. The bright light of heaven never penetrates to the dark caves in which he makes himself a rude home after his own fashion, and weird, wayward life surrounds him on all sides. The diver comes back to the welcome day above, his heart beating high with fearful excitement and his fancy filled to overflowing with quaintest shapes and hideous horrors. Misshapen lumps of quivering flesh, bloated bladders shining in sickly colors, oddly twisted ribbons, with gloating eyes where you least expect them, roll blindly and limbless through the murky waters. Sharp pricks threaten on all sides, long slimy threads slowly and silently wrap themselves around the intruder, and fearful arms of great length, and set with long rows of suckers, stretch eagerly out to catch the welcome prey. Here glassy, colorless eyes stare with dull imbecile light, there deep blue or black eyes glare with almost human sense and unmistakable cunning. And all this world of beings is incessantly at strife; through every submarine bush and thicket glide hosts of fierce, gluttonous robbers. For the calm of the sea is a treacherous rest, and under the deceitful mirror-like peace reigns eternal warfare. Infinite, unquenchable hatred seems to dwell in the cold, unfeeling deep, amid the "things, innumerable, both great and small, that are there."

It must needs be a comfort, therefore, to many denizens of the great deep, to be well protected against the restless spirit of destruction. Happy are the tiny sea-snails, and the countless mussels, who dwell in safe houses of marvellous beauty, presenting to the astonished eye such a variety of turrets and cottages, of staircases and winding passages, of pinnacles and buttresses, as were never dreamt of by human architect. There is an endless variety of stony flowers, now waving to and fro amid the silent currents of the ocean, now rigid and firm forever, when left by the short-lived owner. But all these present but the gorgeous mosaic of the great sub-marine palace; the animal within has little to attract us, and when we draw them up from their dark homes below, it is the house only we value and not the tenant.

Far different is the case with the knight in armor, who leads a strange life, not without humor, in their midst, and blushing bright red for his disgrace adorns our table. His undersized cousin, a mere landlubber, is the familiar crawfish, who dwells in deep miniature caves, next-door neighbor of the bald-tailed water rat, beneath the overhanging network of willow roots and elder bushes on the banks of little streams and brooks. He also is always armed cap-à-pie; his helmet and cuirass in one piece, but the heavy armor below skilfully jointed together and ending in a graceful fin-like rudder. The heavy burden is borne by numerous pairs of stout feet, a very host of legionaries painfully pushing forward the weighty ma-

chine that rests on their broad shoulders. Before him
he bears on high a sharp pair of shears; the first foot has
been changed into a hand, consisting only of a first finger
and thumb, but clever beyond expectation in seizing and
holding whatever it may desire. By the side of the bold
prick which adorns his nose, as knightly horses wore of
old a steel point above the nostrils, rise the two long,
lithe feelers, and upon two delicate pillars appear the
bright, black balls of his eyes, twinkling and twisting
with ludicrous energy towards all sides. Far down in
his innermost recesses he hides a precious stone, the precious gift of Æsculapius, resembling, with the aid of a
lively imagination, a human eye, and endowed with magic
powers. The common people, especially the lower Rus-
sians, still use these so-called stones for many medicinal
purposes, and gather thousands of poor crawfish on the
banks of the Volga, to die a miserable, slow death in the
burning sun, merely to extract from them the highly
prized " white eyes."

Thus strangely accoutred and formidably armed, the
hermit broods all day long in his dark home, a dreamer
in every sense of the word, and a child of dim twilight;
for when night begins to lay her dark mantle upon the
earth, he sallies forth, and, in spite of his weighty armor
and his ungainly shape, he swims about swiftly and
catches many a frog or sleeping fish. Even the water-
snail, in its firm, well-secured house, falls an easy prey
to the great warrior. But, alas! he prefers the dead

body of an animal to all other dainties, and where a poor pike has died a natural death, or a trout has been left wounded sorely by a heron, a whole host of crawfish are soon seen to revel in the feast. Nor do they spare one another, and like many a savage tribe of Africa they also relieve the sick and the aged of their race from the troubles of life by despatching them speedily.

Little valued in this country, the crawfish is looked upon as a dainty dish on the continent of Europe, and hundreds are caught during the bonny month of May, on every creek and every river. But it is not easy to take hold of him; he slips treacherously between your fingers, and if you seize him by one of the claws, he gives it up heroically, like another Scævola, and flees backward into his home. The unwieldy body seems all of a sudden endowed with marvellous agility; he bends the broad tail like a well-tempered spring under the body, and beats with it the water so powerfully that he darts through it like an arrow. His ear is prominent and powerful. The crawfish is the lowest animal endowed with a distinct organ of hearing, and careful observers insist upon it that he is a lover of music. More certain is his love of light, for he is generally caught at night by means of a burning torch; and still more susceptibility does he show for the electric powers of nature, for when a thunder-storm breaks out, he rushes forth from his safe retreat and rages wildly about, as if he feared the end of the world, and vainly sought for a place of refuge.

But what is this pigmy after all to the giant cousin in distant ocean? The lobster is a true knight in armor, fully equipped and of colossal proportions. Some have been found nearly a yard long, true mammoths of long gone-by days, with enormous rods for feelers, and feet covered with knotted hair, while on their broad back a close carpet of mosses and mushrooms had clad the ancient ruin, and snails had found a safe home!

His armor shines like blue steel, unless he should have found a home on rocks strongly empregnated with copper, when his new coat assumes the livery of the sea and changes to green. It is one of the mysteries connected with this strange animal, that he turns bright red when boiled; for no satisfactory explanation has yet been found of the change, and it is not even known whether the new color is the result of a mechanical or a chemical process. Painters love him, therefore, and there are few still-life pictures of the Dutch school that have not a lobster in the fore-ground, now blue amid bright-colored flowers and vegetables in the centre of a market, and now brilliant red by the side of a sparkling glass of wine, and crowned with finely contrasting parsley.

His home lies far down at great depth in the briny waters near rocky coasts, from which he rises only occasionally to lay his eggs and provide for his posterity; for the crustacea have, almost all, curious fancies about that time; the nautilographus fastens himself to the back of a sea-turtle and travels on this safe conveyance through the wide world, while another crab sets forth alone and often wanders over enormous distances. It may be remembered that one of these eccentric creatures was picked up by Columbus in the open sea, when he was yet eighteen miles from land, and gave new courage to his despairing crew, as they saw in the encounter a sign that the new continent was within reach. So true is it that the smallest of beings in His hand may gain an unforeseen influence on the gravest events that regulate the welfare of mankind. He produces eggs, the so-called berries which enrich our lobster salad, and carries them for a while attached to finger-like projections on the lower surface of his tail. These receptacles, in the male animal but short and imperfectly developed, are in the female quite large and full of clusters of eggs during the season. Hence the fishermen know at a glance the sex of their prey, and value their capture accordingly. But the lobster represents in the sea the marsupial tribes of the land, and as the kangaroo carries her newly-born young in her pouch, the lobster also keeps the tender off-spring for a while under the secure shelter of the broad rudder-like tail. The eggs are amazing by their number, for not less than twelve thousand have been counted in a single female, and yet the enemies are so numerous and so voracious that without this gift of parental affection bestowed even upon a creature so low in the scale of beings, the race would have long since become extinct. They love their young very manifestly; for the younger Buckland tells us, in his Curiosities, that fishermen of Cornwall frequently see lobsters surrounded by their young, even when already over six inches in length. Or the mother would be noticed lying with her head peeping from under a rock, with her large claws extended, while the young ones were playing merrily between them; when danger approached, the old one would rattle her claws and the young ones at once seek shelter under the rock.

The numerous legs are but feeble and barely able to drag the heavy armor slowly over the ground, but far down in his own element, the lobster glides rapidly over the rocks and reefs. His many-linked tail, well-jointed and yet extremely pliant and agile, possesses great power, and with a single blow a full-grown lobster will dart to a distance of fifty feet and instantly escape pursuit. To aid him in his movements, he is endowed with a marvellous instinct, which enables him, though moving backward, always to hit exactly the entrance to his little cave, distant as it may be, and barely large enough to admit his body.

The lobster, however, is not only larger and stronger than his cousin on shore, but he shows also superior faculties. Far from all tendency to cannibalism, he lives in friendly union with his brethren, and often joins a merry company on a common excursion to distant seas. When in his warfare strength does not avail him, he resorts to stratagems and shows great cunning. Thus he never despairs of conquering the stubborn resistance of shell-fish; patiently he lies in wait for hours and hours, until the poor animal, lured into security, timidly opens the shell. Quick as lightning, he shoots up and places a tiny pebble between the two valves, and the oyster surrenders. Nor is he less susceptible to electricity; for the thunder of the clouds or the roar of cannon affects him in his remotest caverns, so that he wildly rushes out and in his terrible fright casts off his claws. Freebooters are reported to have taken advantage of this idiosyncrasy, to threaten poor Norwegian fishermen with the firing of their guns, if they were not willing to share with them their loads of lobsters.

Their claws are the result of a most ingenious and yet marvellously simple device of nature. The first foot is inserted sideways in the second, and thus forms a kind of shears; the changed foot becomes, of course, unfit for locomotion, but in return extremely useful in seizing the prey, in resisting an attack, and overcoming an enemy. With them the common lobster also carries food to its mouth, and skilfully skims the water to catch all particles of food that may float on the surface. These claws, as well as the feet, can be thrown away under the influence of fright, or be lost in the heat of the combat, without causing pain or special discomfort. The mutilated animal runs away on the remaining legs as if nothing had happened, and soon sees a new limb replace the lost one; nor does the cast-off claw seem to be much missed until a substitute has grown out again. The latter, however, is never the same size as the old one, and hence lobsters are so frequently found with one claw much larger than the other. Here, also, we cannot help admiring the benevolent wisdom which has endowed animals so constantly in danger of having their limbs snapped off by countless enemies, and yet so entirely dependent on them, with the power of reproduction. The time for the latter is not always the same; it depends much on the warmth of the season and the supply of food, as well as on the part which has been lost; the tail is never replaced, and the animal that has lost it dies without fail.

But by far the most interesting feature in the life of the lobster is the change of his armor. This coat of mail is of one piece, and consequently incapable of extension or alteration; hence the lobster, having once grown up to the size of his house, could never hope to grow beyond the tight uncomfortable garment of his youth, if an all-wise Providence had not provided a way, by which he may change his armor at least once a year. At the proper season, generally toward the end of spring, when food is plentiful, the knight begins to feel ill at ease in his close armor, and seeks some dark cleft in the rocks, or other dark place of retreat, where he may undergo, in seclusion and security, a change that exposes him to great suffering in body, and much danger from abroad. Here he begins to agitate his limbs, to move in violent contortions, and to swell out his body. After a little while the shell bursts, like the cracked bark of a tree, splitting exactly down the centre of the head portion, so that a slight pull would tear the two parts asunder. The shell then comes off in two halves, exactly as the cuirass of a modern cuirassier or a Horse Guard's man ; then follows more pulling and jerking till the legs also come out, and at last the tail even follows the example, and slips out quietly, like a hand withdrawn from a glove. But the process is not so easy with the claws, broad as they are at the end, and very narrow at the points of juncture. Fortunately the flesh of the animal has become, at this season, quite soft, and as elastic as India-rubber; by long continued efforts the broad hand is drawn slowly through the narrow wristband and soon spreads out again into its former shape. Sometimes, to be sure, an impatient creature pulls too suddenly, or too violently, and the hand remains in the glove, but they seem to mind the loss but little. When the whole operation is over, which generally takes three days, the knight sinks into a state of utter exhaustion; the limbs are so soft and limber that they bend like pieces of wet paper, and only on the back the flesh has retained some firmness. Yet not a particle is wanting ; every delicate feeler has shed its outer coating ; the eye has lost its covering, and even the stomach has cast out its lining membrane. A shining secretion moistens the whole body and helps during the painful process.

Now the poor animal wants rest, and above all, a place of perfect security; for he is a knight in armor no longer, but utterly helpless and defenceless. After a few days, however, the outer covering begins to harden, and in a short time the happy lobster, about one-fifth of his size larger than before, enjoys the bliss of being young and beautiful once more, and feels, in his bright and strong armor, no doubt, as happy as a lobster well can feel. What would man not give for this most enviable power to renew the outer coat from time to time, and even to restore the stomach to primitive power and freshness!

Now he sallies forth, once more, armed cap-à-pie, and ready to encounter all adversaries and to overcome all enemies. He is starved, and fearful is the havoc which he makes among small fry, and all weaker animals that can serve him for food. Now also he shows the most marked evidences of the acute sense of smell with which he has been endowed. Nine miles out at sea, says Buckland, off Lyme-Regis, in Dorsetshire, there is a ledge of chalk rocks which abounds in lobsters, and here this remarkable instinct has been most accurately observed. They will smell a putrid object, down in the water, at a distance of more than a hundred yards, and when a shipwreck occurs, it becomes at once known to the horrid epicures. A vessel thus once perished off the island of

Portland and many persons were drowned ; soon afterwards a great number of prawns and lobsters were noticed
in those waters, and hundreds of the latter were caught.
The good people of Weymouth refused to buy them, as
they were suspected, and very justly, to have fed on the
bodies of the drowned people, but they did not hesitate
to send every one of them to London, for the benefit of
those who knew nothing of their sad history !

A PINCH OF SALT.

"Salt is good."—St. Mark ix. 50.

THE servant of the great chemist, Berzelius, was once
approached by one of his countrymen with the
question, "What is that chemistry by which they say
your master has made himself so famous?" "I will tell
you," was the ready answer. "First, I have to fetch all
manner of things in large vessels; then he pours them
into bottles, and at last into quite small phials; when he
has done that, he pours them all once more into two big
buckets, which I carry out and empty into the river.
That is chemistry."

The popular idea of the science is not much clearer in
our day. The name conjures up, in many minds, a large
laboratory, with quaint retorts and vile smells, or at best
a huge factory, sending forth clouds of disgusting smoke.
In many a lively imagination the chemist is still surrounded by stuffed monsters and bottled infants, after the
manner of Hogarth's admirable etching, and his labors
are looked upon with timid admiration and doubtful

7

wonder; for the alchemist has not yet entirely faded away into a myth, and the Black Art has still his votaries in our midst. Few among us are really aware how deeply and practically the chemist's science affects our daily life and contributes to our happiness upon earth.

And yet he has a duty to perform which ranks but little below the very highest that falls to the lot of man here below. He is the self-appointed guardian of the indestructible part of our globe. Man glories in his absolute sway over all Nature, whose gifts he employs for his pleasure, and whose creatures he treats as his vassals. But his dominion is of short duration, and soon Nature resumes her own sway again, unimpeded by his hand. He wrests massive rocks from her bosom, and tears gigantic trees from their ancient homes, and changes them into houses and palaces and ships; he digs into the bowels of the earth, and fashions the hidden treasures into bright ornaments and useful tools, or he transforms even the worthless sand and the shapeless clay into costly wares of brilliant splendor. But a few years pass away, and his beautiful handiwork changes in shape and in color; a century more, and they crumble into dust. His magnificent temples, his lofty walls, his graceful bridges, his proud monuments that were to give immortality to his name and his deeds—they all succumb, sooner or later, to the silent but unfailing efforts of Nature to reclaim her own children. What the waters of the ocean and the winds of heaven have left undestroyed, falls under the unseen attacks of frost and rain and heat. As worms feed under the green turf on his body, fearfully and wonderfully made though it was, so tiny lichens and minute mosses consume, little by little, his obelisks and his pyramids. Diminutive seeds, flying unseen through the air, come and nestle in the cracks and crevices of his castles and palaces, and strike their frail roots in the rents of his massive walls, while treacherous ivy sends his tendrils into every cleft of the ruin. Insects, creeping about by night, undermine the foundations of colossal structures, and animal life teems ere long among the *débris* of his loftiest edifices. The trees he has planted and the animals he has raised, return to the dust from whence they sprang; the wood he has carved with cunning craft, decays into impalpable powder; the metals he has wrought into shapes of wondrous beauty, are eaten up by rust; and the very stones he has piled up in lofty structures, are consumed by wind and weather.

And whither go all these fading, fleeting elements, which thus continually pass from his sight, and return, as he calls it, to the bosom of their mother Nature? The chemist alone can answer the question; for he alone watches them forever, and never for a moment fails to trace them to their new home, though they assume, with Protean power, a thousand new forms, and defy him, for a time, by their incessant and marvellous changes.

But his power is greater yet. For this knowledge of the eternal duration of the elements in nature endows him with a power that might almost be called creative ; for though he may not absolutely produce them out of naught, like the one great Creator above us, he can at least make them assume the form which he wishes. He can take the dust, that seems worthless, and endow it with priceless value ; he can gather impalpable powder and hardly perceptible vapor, and bid them combine in a form that shall rejoice our eye by its beauty, and prove itself a blessing like few others to all mankind. One of the most striking instances of this power is the manner in which his science transforms an unsightly gift of nature into the most precious boon that man receives at her hands—a little gray substance into a priceless crystal, far more valuable than all the most costly jewels he possesses, and indispensable to his very existence upon earth.

This precious treasure is a little square-fashioned grain, of gray color, born far down in the darkest recesses ⸱of the earth, in times when fierce fires raged below ; and there it has lain for thousands of years, along with countless little grains of like shape, never seeing a beautiful flower by its side or hearing the sweet notes of a bird as it sings of spring and budding love. Its ancestors were two strange beings, that have but quite lately become known to us : a metal with a silver sheen and a gas of yellowish-green color. The former is perhaps the oddest of its kind. Other metals are heavy and hard ; this is so light that it swims on the water, and so soft that it can be cut with a knife and kneaded with the hand. Other metals resist all impressions from without; this is so yielding that if exposed for a little while to damp air, it oxydizes quickly and changes into a white powder! While its near cousins, gold and silver, sink quickly to the bottom of a vessel filled with water, Sodium, on the contrary, floats like a very gnome of the mountains; and the little silvery globes, in which it is ordinarily seen, swim merrily for a while on the surface. But after a few moments, they begin to glow and to shine like liquid fire, and now perform a dance so weird and wild that it startles us by its strange, fantastic figures. The smooth surface of the water becomes the well-polished floor of a ball-room, on which the bright pearls of shining metal perform their quaint dance like enchanted princesses dressed in silver robes. If you attempt to hold the lovely little dames by force, they know at once how to escape from your violence and to regain their liberty. A beautiful bluish flame begins to surround the little globules, and a few moments after the metal has vanished. No trace is left, and only the peculiar smell of the water betrays their secret: they have sought refuge in the friendly element, and water, the stanch enemy of fire from of old and ever ready to conquer it by its own power, has been forced by the little magicians to burn, for a little while, in a bright, flickering flame, before it could afford them a new home and a safe shelter.

Nor is the other ancestor of the tiny grain less odd in
its nature. While every other substance on earth has
some shape and form of its own, by which it may be
known, and even water, though ever changing and restless, never fails to fashion itself in lovely globules, Chlorine has no form or substance of its own.   Like a prince
of the air, it passes unheeded through the atmosphere,
visible only as a faint, yellowish-green vapor.   You catch
it and imprison it in a glass, you compress it by all the
means at your disposal with terrible force, and at last it
comes down reluctantly, in the shape of a bubbling liquid.
But relieve it for a moment of the enormous pressure, and
it rises instantly again as a vapor, and escapes from your
grasp.   Unfit to be breathed by man or beast, whom it
would smother in a few moments, it yet is not merely fatal to life, but has its good ·use in the wonderful household of nature, where every atom finds its noble vocation,
and serves its great though often unseen purpose to the
glory of the Most High.   Chlorine has been endowed
with a truly wonderful power of combining with all other
elements without exception, and hence becomes of vast
importance to the chemist and the manufacturer.   But
it serves us must faithfully where danger threatens us
most nearly, and carries off, with unfailing fidelity, the
death-bringing gases of wells and neglected cellars, and
purifies our sick-rooms and hospitals.

These two strange beings, the flitting gas with its repulsive color and fatal breath, and the quaint metal whose

merry dance forces water to turn into fire, seek each other, throughout nature, with unceasing longing. And yet, whenever they meet, they embrace each other only for a moment, and for their own destruction. The bright silvery substance has no sooner been breathed upon by the foul gas, than it vanishes as if by magic, and all that remains of the two is a tiny crystal of white color and silvery sheen. You examine it closely, and you find that it resembles a hollow cube; every minute particle of the grain is clear and transparent, like the most perfect of crystals, and it is only when many are lying close to each other that the broken rays of light give them a pure, white hue. The poisonous powder of chlorine and the fiery nature of sodium have utterly disappeared, and in their stead man is presented by his beneficent mother Nature with a little grain of salt, without which his life would be a burden, and happiness upon earth forever out of question!

Fortunately, Nature is as bountiful as she is wise, and hence the indispensable grain of salt is provided by her with such a lavish hand, that it may be found in immense quantities all over the earth. The land hides it in its dark caves, and holds it forth in large shining masses on the surface; and the sea is filled with it, from the topmost wave to the bottomless abyss. For the unsightly grain, little noticed by careless man, and taken as a matter of course by most of us, is the great guardian of health throughout our world; without it the waters of the earth would soon stink with corruption and all flesh would be foul with decay; without it the plants would no longer deck the land with their beauty, and man would die a death of misery and unspeakable horror. Hence the mercy of the Creator has scattered it broad-cast over our domain, and we have but to stretch out our hand to gather the precious gift from on high.

For away, in Eastern Europe, the traveller comes up-on a long, low range of hills, stretching from east to west, which enclose, with their soft outlines and well-wooded slopes, a lovely valley, dotted here and there with smaller hills and little knolls. A cluster of low but well-kept houses lies toward the opening from which he approaches the plain, and the eye wanders freely beyond them into distant lowlands. It is a busy scene to which he comes, and men are moving briskly about through the narrow streets and the countless paths that lead over the common. They wear a strange costume of sombre black, and have thick leather aprons tied on behind instead of in front; but they look cheerful and happy, and many a merry song and sweet carol is heard far and near. The traveller engages one of these men, who all greet him with a pious wish for his soul's welfare, to show him the way into the mysterious world below, of which he has heard much; and soon he finds himself arrayed in a white blouse and black velvet cap, such as are kept ready for visitors, at the mouth of a shaft which seems to lead down to the very bowels of the earth. When his eye has be-

come somewhat better accustomed to the dim light of
the candle stuck in his hat, he notices that wooden rails
are laid all the way down the gently inclined plane; and
he is invited to mount a wooden contrivance, wondrously
like the hobby-horse of our happy childhood. The miner
sits down before him; the horse— a sausage, it is called
in local parlance—starts with alarming swiftness on the
smooth, oiled rails, and his right hand, armed with a
stout, leathern gauntlet, grasps frantically the rope that
runs along the wall, to check the painful velocity. At
last the two horsemen are stopped, by reaching a piece
of level ground, and the traveller finds himself in a vast,
subterranean corridor, cut out of the live salt. Huge
blocks of the precious material are lying about, some
colorless, some shining in beautiful though subdued blue;
the roof rises high above him, and looks gray and grim
in the dim light, and on his right the vaulted ceiling rests
on gigantic pillars, in which each tiny grain shines
brightly and sparkles as the light falls upon it; and yet
they all hold so firmly to each other that there is no
danger of their ever giving way and proving faithless to
their trust. A little further on the miners are hard at
work; they attack the mountain-side by cutting out immense blocks in the shape of huge casks; then water is
poured down the furrows and allowed to remain standing there a few days, so as to soften the rock; at the
proper time wedges are driven in, which soon swell in the
water and blast out, as it were, without further help from
7*

human hands, the great blocks in the desired form. As the traveller wanders on through the long dark passages, with statues in niches and holy images at the corners, he passes large vaulted rooms, dark caves, and huge recesses, that seem to have no end, and at times he comes upon stairs, cut in the rock, which he has to descend cautiously, so smooth and slippery is the material of which they are formed. Every now and then he sees, at a distance, a bell-shaped shaft, from the top of which hangs a frail ladder, free in the air, swaying and swinging to and fro with the cold currents that blow here perpetually ; and he looks with wonder and fear at the poor miner, who trembles and crosses himself piously, as he sets foot on the slim rounds and descends slowly into the apparently unfathomable darkness below. All of a sudden he sees bright lights before him, and, dazzled and surprised, he enters a vast cathedral, the walls of which shine and shimmer all around in fanciful, flitting lights, as the light of torches and candles fall upon the bright masses of salt ; there is the altar with its colossal cross, and at the side the organ and choir ; here also statues and images abound on all sides, and even human worshippers, kneeling down in silent adoration, are cut out in the yielding material. He has little relish, perhaps, for the vast ball-room, with its orchestra on high and tis brilliant chandeliers, glittering and glistening like the fairest of crystals- and bed-chambers with mocking couches ; for the whole upper world is repeated here below in grotesque caricature.

Gradually the passages become lower; the ceiling sinks more and more on the left, and at last the traveller is forced to bend, until he fairly creeps along on all fours. But suddenly he sees before him a fairy scene: dark waters, sparkling bright in the light of torches fastened to the glistening walls. Like a vast black mirror, the subterranean lake, silent and motionless, stretches far into the endless darkness. Never has wing of bird dipped its feathers into the mysterious water; never has a breath of air ruffled its placid, patient surface. Like walls of iron, the rocks of salt rise all around in grim solemnity, and hold the restless element bound in eternal silence and peace. The scene is beautiful, and yet fearful in its utter loneliness and death-like stillness.

A few shells and *débris* of marine-plants are found on the banks of the black tarn, but they belong to generations as old as the Deluge. No life has ever been known to grace the lake. Only ages and ages' ago, when the waters that now rest deep below the world of men, were purling merrily down the mountain-side, they bore with them the tiny houses of friendly animals; and in their wanderings through the hidden depths of the earth, carried them with them to their silent home. At the further end, to which the traveller is rowed in a crazy punt, a little chapel rises, unpretending and unhonored, and yet of great import. It is devoted to the memory of the pious wife of one of Poland's early kings, to whom Heaven vouchsafed, in 1252, the boon of bestowing the knowledge of these wondrous treasures on her impoverished subjects. She was afar off in Hungary, the legend says, and hearing there of the fearful suffering of her native land, she was ordered, by her patron-saint, to cast a precious ring, which she most valued of all her trinkets, into a deep well. She did it in simple faith, and, when she returned to her home at the foot of the Carpathian Mountains, some peasants brought her a piece of rock-salt, believing it to be a costly jewel. It was of no value in itself, but, oh wonder! in the heart of the transparent mass her ring lay imbedded. She understood the revelation from on high—ordered search to be made for more of the shining substance, and thus were discovered the great mines of Wieliczka, which have ever since been a source of greater wealth than the richest mines of gold or diamonds.

Beyond the little chapel the work begins once more, and miners are seen busy loosening vast lumps of salt from the parent mass, blasting the less pure material with powder, and cutting out the more valuable blocks carefully with chisel and chipping-knife. Others harness the twelve horses that are kept below and have never seen the light of heaven, to rude sledges, on which the blocks are drawn to the foot of the shafts, that lead up to the world above; while still others are opening new passages or propping up dangerous places with large wooden pillars. With a feeling of pity for their hard work and thankfulness for the boon they bestow upon mankind, the traveller passes them, returning their friendly greeting, and gladly beholds once more, as he rises to the top of the shaft, the bright light of day and the fresh air of the earth above.

Thus the salt is found crystallized in large beds and boulders, stowed away between layers of clay and lime-stone, in more or less regular shapes, and then called rock-salt. Nearly every part of our globe is endowed with vast deposits of the kind. Bergen in Norway, and Cardona in Spain, vie with each other in the abundance of their supply. In the latter place, a huge mountain of almost pure salt rises clear and sheer from the plain, the whole mass shining brilliantly like a glacier in the sunlight, or glittering in a thousand hues and shades, when day fades away. The salt here is so hard that it has to be blasted, like real rock, with gunpowder, and the chips are worked up by skilful hands into snuff-boxes, crosses, and rings. Norwich, in England, boasts of a field of salt more than seventy-five miles long; Salzburg proudly bears the name of its staple product; and Mexico and Persia, the East and the West, are all full of ample supplies, which, by God's providence, have been laid up in store for many generations to come.

Not in all parts of the world, however, is salt found so pure as to be fit for immediate consumption. Generally it is mixed up with clay and sand, and then has to be purified by the aid of water. Man leads the purifying element down to the beds of rock-salt, allows it to dissolve as much as it is capable of holding, and then raises it, by vast pump-works, once more to the surface of the earth. In vast kettles and pans, beneath which huge fires burn day and night, the brine is then evaporated, and white crystals of salt remain, pure and unadulterated, at the bottom and on the sides of the vessels.

In other regions Nature is even more liberal, and saves man the necessity of leading the water down to the depths in which salt is hidden. Large rivers beneath the ground are led, by the hand that holds the earth in its grasp, over extensive deposits of salt, and then break forth as saline springs at the side of the mountain. Thus there is near Minden, in Prussia, a well nearly two thousand feet deep, which holds a water, the temperature of which exceeds 25° Réaumur, and which is, below, continually dissolving large blocks of salt, in order to gush forth above and bring the precious gift up to the surface. Germany boasts of not less than eighty such valuable springs; and our own country is most richly endowed in like manner, so that the two States of New York and Virginia could supply, if need be, the whole of the Union with the salt they require.

Brilliant as it appears in the shape of rock-salt, and pleasing as are the waters of saline springs to the eye, salt yet presents itself, at times, under an aspect much less inviting. No words can describe the horror of the vast salt-plains, which here and there interrupt the beautiful carpet that covers the surface of our earth.

Thus there is a vast district in South America, extending over more than twenty thousand square miles, which forms one enormous group of desolate mountains intersected with vast deserts, saline swamps, and dried-up salt-lakes. Currents of hot air meet here from all parts of the compass, and with such vehemence and persistent fury, as they rise incessantly from the heated, steaming soil, that no clouds can be formed and no rain can fall from the ever-serene sky.

Even more fearful yet is an endless, lifeless plain in the heart of Persia, so sterile and accursed than even saline plants do not thrive here; but the salt itself, as if in bitter mockery, fashions its crystals in the form of stems and stalks, and covers the steppe with a carpet of unique vegetation, glittering and glistening like an enchanted prairie in the dazzling light of the Eastern sun. In the rare places, where the thick crust is broken and vegetation is favored by night-dews, a few straggling herbs and grasses appear; but they are saturated with salt and soda, the sap tastes bitter and salty, and stalks and leaves alike are covered with a thick incrustation of salt, as if with impalpable powder. They afford no nutriment to the herds, and soon give away again to the genuine salt-desert, where shepherd and flock alike find their death. For here a light, loose sand rules supreme, now treacherously quiet, but sure to engulf the heedless herdsman who puts his foot on the glistening surface, and is swiftly sucked in by the tricky soil; and now rising in large, deep-red clouds, which fill the valleys and level the ridges, till every landmark is effaced, and the whole vast region resembles a petrified ocean of blood-red waters.

Who can describe the bitter, mournful disappointment of the thirsty traveller, who sees, at last, afar off, the welcome glittering of waters, and hastens, with renewed vigor and high hopes, toward the enchanted spot? Enchanted, indeed! For as he approaches, the fairy spectacle strikes him with wonder and sad misgivings. In the midst of the brown, desolate plain, a vast level sheet of pure white stretches far and near; he draws nearer, with faltering, doubtful step, and sees, at last, to his horror and dismay, that what he fancied a basin of cool, refreshing water, is nothing more than a white crust of salt. Or, it may be, he descends, with eager expectation, the steps hewn in the precipitous walls of an ancient crater in South America, of which Darwin tells us, in order to reach the little circular lake, embosomed among rugged fields of lava, and fringed with a border of bright-green, succulent plants. As he looks down from the immense tuft crater, he sees the water clearly, and fancies his ear even discerns the pleasant splash against the modest beach; but when he reaches the lake and dips his parched lips into the liquid, he draws back with dismay; for it is bitter and brackish, and unfit for the use of man. Other travellers tell us of the sad fate of black slaves who work in the salt-plains of the Sahara, collecting the salt from the surface, hundreds of miles away from the nearest oasis, and sure to perish by hunger and thirst, if the caravan that is to bring them food and water should lose its way in the desert or fall into the hands of merciless robbers.

Even Europe is not free from these unfortunate places, which seem to bear the curse of Sodom and Gomorrah, and have become what Zephaniah threatens, " a breeding of nettles, and salt-pits, and a perpetual desolation." Here nothing grows but impoverished looking plants, with pale, bluish-green color and faded blossoms, which give to the region an air of overwhelming monotony and ghastly sterility. The burning rays of the sun are mercilessly reflected from the white crust of salt, which covers the soil, with such fierceness, that the eyes are unable to bear the unearthly splendor, and the soil opens here and there in huge cracks and crevices, burned, as it is, to the core, and but rarely refreshed by scanty rain or nightly dew.

How did these desolate lakes originate, and whence come the bubbling springs which so industriously bring up to their master the salt he needs for his life ? The question, for a long time, defied the wisest among men; but modern science has solved the riddle, at least with regard to the latter. We know now that the water that comes in the shape of snows and rains from the skies, and of the dew distilled near the surface, slowly but surely finds its way, through the porous crust of the earth,

down to the interior of mountains and far below the
level of plains. It stops not till it meets with a layer of
firm rock, which prevents it from sinking still lower;
and here, on the unyielding stone, it forms, gradually,
subterranean lakes; the waters are not at rest yet, but
silently and steadily keep on, dissolving all that they
can reach around them, and thus they become saturated,
now with sulphur or salt, and now with minerals of
every kind. When man discovers such a spot, he sinks
a shaft to the basin below, and at once the waters, relieved of the pressure, leap up in wild joy at their return
to the bright light from which they came, and rise as
high, once more, as the place where they first entered
the earth. Science tells us, of course, that there must
ever be found, near such springs, large beds of salt;
and this has led, of late, to most valuable discoveries of
immense deposits in Germany and in France.

The origin of extensive surface-beds of salt, such as
are found in the vast steppes near the Caspian Sea and
the Aral, high above the surrounding country and far
beyond the reach of supplies from a distance, is less
clearly understood. Some believe that they are the
beds of ancient oceans, from which the water has gradually evaporated, leaving nothing but the bare bright
crystal behind. This explanation may apply to the
Siberian salt-plains, which, like the Sahara, were no
doubt once the bottoms of great oceans, drained by
some fearful upheaving of the ground or the breaking

down of gigantic walls, which formerly held in the waters
of the enormous inland lakes.   But with regard to others,

> None can reply—all seems eternal now.
> The wilderness has a mysterious tongue,
> Which teaches awful doubt.

Others think that the salt, which now glistens on the
surface, once lay buried far below, and was raised, by
volcanic upheavings and fiery eruptions, in the shape of
boiling brine; the waters then evaporated, or were carried by rivers into the sea, and the salt remained spread
out on the low bottom of the steppes.   But this theory
would hardly account for the strange fact, that the salt
on these immense plains actually grows there; it is no
sooner removed by the hand of man, than it begins to
reappear, and ere long the crust is close and compact
once more.   This is the case with the terrible Desert of
Dankali in Abyssinia, where, for four days' journey,
nothing is seen but a rank vegetation of apparent plants,
with their stems and leaf-stalks all of salt, and where no
effort to clear the soil ever makes the slightest impression.
The same has been observed near the Salt Lake of Utah
and on the banks of the Mingo Lake in Texas, where the
crust of salt is so thick that it can be removed in large
blocks, and yet no diminution is ever observed.

Where neither masses of rock-salt, nor waters holding
large quantities of salt, provide for the wants of man, he
knows how to force the very plants that delight, like
him, in the precious boon of nature, to furnish him all he

desires. For it is not the miner alone who goes down
into the deep of the earth to search for salt, but plants
also send down their roots, draw up the salt water, and
deposit the proceeds in beautiful crystals in their cells.
There are few plants, altogether, which do not contain in
their delicate tissues a certain quantity of salt, especially
in the stems and the branches, and leave it behind in
their ashes, when they are burned. Some cereals require
it, therefore, for their satisfactory growth, and much salt
is sown on the broad lands of England and the fields of
China; others, like asparagus and flax, do not thrive at
all without such aid. But the growth which surrounds
salt-springs and the plants that love to dwell on the sea-
shore, delight in the little grains; even the lofty cocos-
palm sends its large oval fruit adrift, to seek some briny
strand, where it may find a rich soil and abundance of
salt; and the careful husbandman of those regions, when
planting the nut that is to give him his daily bread,
drops a handful of salt into the hole, to which he con-
fides the gigantic seed-corn.

Here and there, in favored lands, you see a vast,
marshy meadow, spread out in beautiful luxuriance before your eye, dotted with pretty copses of elders and
willows. Close by one of these groups of low, spreading trees, where the soil almost imperceptibly rises into a
little knoll, there gushes forth a clear, powerful spring,
and forms, at its very birth, a large, circular basin, filled
with transparent water. A rivulet runs from it slowly

but steadily, wanders, as if enjoying the luxury of leisure, through level meadows, saturating the porous soil on
the right and the left, and at last falls, at the edge of the
high table-land, with merry laughter, into the lower
plain, to bring its modest tribute to the large river below.

There are other meadows scattered over the plateau,
but not one of them can boast of the bright flowers and
waving grasses which bud and blossom forth in unwonted richness.   Thousands of purple asters peep out
with their bright eyes, set in golden yellow, from the
midst of dense clumps of reeds; luxuriant plantains overshadow a host of minor plants of strange and uncouth
appearance, and a variety of glaux spreads all around a
deep-green carpet, strewn with an abundance of small
white flowers.   Further on, a quaint salicornia appears,
in large patches; its long-linked stem looks as if it would
burst, filled, as it seems, to overflowing with exuberant
sap, and in the axes between the branches, lurk countless
diminutive blossoms of bright yellow.   Even the grasses
and reeds which cover the marshy ground, when more
closely examined, prove to be entirely different from all
that grow on adjoining lands.

The flocks of birds who have left their homes in the
far north, and now, with swift wings, move southward to
more genial climes, might fancy they beheld here, once
more, the shores on which they last sought rest and repose.   For here are the same flowers which they saw

there, near the downs; the same lowly herbs that love to be bathed daily in the briny waters, and the same reeds that grow there within reach of the unfailing tides. For it is a salt-spring which here wells up, and unable, at once, to reach the lowlands by any other outlet, has here formed a lake, and furnished food to an exuberant vegetation.

It is from these saline plants, growing now near the shores of the ocean, and now far inland around merry springs, that large provisions of salt are won by the aid of fire. The soda, or barile of commerce, comes almost exclusively from the ashes of the saltwort, a plant of grayish green color, with stems a foot long, thickly set with prickly hair, and with uncouth, swollen-looking leaves, ending in sharp, pointed thorns. The Arabs hardly knew what a blessing they bestowed upon mankind, when, upon settling in Spain, they brought with them not only their merino sheep, their cotton and sugar-cane, but also the unsightly saltwort, from which they already knew how to obtain the soda of our day.

Another salt-plant, the leafless glasswort, is eaten as a salad in England and the whole north of Europe; but the most curious of them all is perhaps the variety known to our green-houses as the ice-plant. This strange-looking plant is a treasure to the inhabitants of the Canary Islands, who raise it in large fields, pull it up when ready for use, burn it, and drive a most profitable trade with the soda they obtain from the ashes.

It is, however, not the water only which gives us salt, but we owe it also, at times, to the benevolence of fire. For although the beautiful crystals do not become volatile till they are heated to a white glow, they are still not unfrequently found among the strange medley of substances thrown out by volcanoes. After an eruption, the cracks and crevices of Mount Vesuvius are often covered with a thick crust of salt, and the surface of petrified streams of lava appears, at times, from the same cause, as if thickly strewn with white powder. In 1822, the salt cropped out in such very large masses, that the greedy Government of Naples laid an embargo on the treasure, and obtained, through its own workmen, blocks of twenty-four feet square from the vicinity of the crater. The same takes place occasionally at the foot of Mount Hecla, in Iceland, and the industrious peasants carry whole wagon-loads to their fields and their houses.

Such is the history and home of the .precious little grain, which the world, from the beginning, has looked upon with a feeling akin to awe and reverence. For while deeply grateful to the Giver of every good and perfect gift for the tiny crystal, on which life itself is dependent, men have ever felt that it was endowed also with a dread power of final destruction. The ancients had no doubt that salt was a direct gift of the gods, and hence they joined it, symbolically, to every sacrifice offered on holy altars ; and Moses ordained that " every oblation of thy meat shalt thou season with salt : neither shalt thou suffer the salt of the covenant of thy God to be lacking from the meat-offering : with all thine offering thou shalt offer salt." The Aztecs of Mexico had a special goddess presiding over the use of the indispensable condiment; the Chinese celebrate, to this day, an annual feast in honor of him who first introduced it into general use ; and the old Egyptians, when they performed the rites of their great festival in honor of Neith, the mother of life, filled the lamps of their temples with salt as well as oil.

Miraculous powers, also, seem to have been attributed to salt, from olden times ; for the Hebrews used to rub new-born children with it, partly from a belief, sanctioned by Galen, that this hardened and strengthened their skin, and partly from faith in its special blessing. Hence the prophet Ezekiel reproaches the stubborn people, by saying: "Thou wast not salted at all, nor swaddled at all;" and even the early Christians adhered to the old usage, for they initiated young converts into the mysteries of their faith by placing salt in their mouth, as they did with infants at the time of their baptism.

It was but natural, therefore, that the semi-sacred character of salt should lead soon to its being used in connection with treaties and compacts to render them more binding. The Old Testament is full of allusions to this ancient usage, and Moses already speaks of " a salt-covenant forever before the Lord unto thee and unto thy seed with thee." Its power to protect against corruption lent its symbolic force to stipulations even among infidels, and few such compacts were made without a plate of salt being placed ready at hand, from which each of the contracting parties eat a few grains, instead of swearing an oath. The Arabs of our day still enter into the most sacred treaty of friendship with each other by pushing a piece of bread, strewn with salt, into each other's mouth, and then call it a " salt-treaty." The ancestral salt-cellar, that played so prominent a part in the household of ancient Romans, was, in like manner, the great symbol of the union that bound the members of a family to each other.

Scarcely less general is, however, the dread which salt inspired by its strange power of destroying the productiveness of the soil; and thus it became, very early, already the symbol of sterility also. Jeremiah cursed Judah, by condemning it " to inhabit the parched places in the wilderness, in a salt-land, and not inhabited; " and the terrible fate of Lot's wife has left the curse vivid in the memory of men. For the same reason, when Abimelech had destroyed the city of Sichem, and rased its walls to the ground, the place where it had stood was sown with salt, not in order to make it sterile, but as a sign that it should remain waste forever. Even the Middle Ages employed the dread symbol; and the great Barbarossa, after taking rebellious Milan, and destroying its beautiful buildings, ordered the plough to be passed over the city, and then salt to be strewn on the

8

spot, leaving only the churches unharmed, "for the greater glory of God."

On the other hand, salt makes "unsavory things" palatable again, as Job already mentions; and hence it soon became usual to speak of it as a symbol of that sagacity which uses apparently worthless matters for a good purpose, and employs words of trifling import in themselves with great effect. This was the first meaning of Attic salt; hence, also, St. Paul writes, "Let your speech be alway with grace, seasoned with salt, that ye may know how to answer every man;" and the Saviour Himself calls His disciples "the salt of the earth," as men by whose instruction and example their brethren are to be taught and saved from condemnation.

All this worship of salt as a divine gift, this veneration of its sacred character, and this dread of its destructive powers, centre, however, in the simple fact, taught by modern chemistry, that salt is the great regulator of the health of the world. Without it the seas would be impure, and the land a desolate scene of destruction; man would not be able to live, and the beasts of the field, with the plants that feed them, would no longer be seen. The little grain of salt at which we hardly glance, is thus of vital importance in the great household of nature. But it shares the fate of all indispensable things by which we are surrounded: habit makes dull the sensibility of our senses, and with it the activity of thought that depends on such impres-

sions. Only what is rare and unusual attracts our attention, though it have but an outside brilliancy and useless
beauty. The sparkling diamond is sure of admiration;
set in bright gold, it is esteemed above all things, and
serves to enhance beauty, to display our wealth, or to
symbolize supreme power. The unattractive twin-sister,
black coal, has to do hard work in the kitchen, the workshop, and the factory, like a true Cinderella; and yet on
coal, and not on the diamond, rests the true wealth of a
nation, the foundation of happiness for countless millions. Thus it is with the tiny grain of salt; rich and
poor see it, day by day, on their table, and enjoy it with
everything they eat and drink, but few ever inquire
whence it came, and what accident or what necessity
brought it there. And yet, let it be missing but for a
single day, and how we would suffer!

We all know that the ocean is salt, and that without
it neither animal nor plant could live in the vast basins
of the earth. But it is less generally known that the
amount of salt in different seas is not the same, but
steadily decreases in the direction from the equator to
the poles. Scoresby tells us that, of European seas, the
Mediterranean holds most, the Baltic least; so that the
fishermen of the north have to send for the salt they
need in preserving their fish, to the more favored regions
of the south, and salt becomes a patron of active trade.
The Atlantic Ocean, again, has more salt than the
Pacific, and the Polar Sea least of all. With the amount

of salt, which makes the water denser, and thus better able to bear heavy vessels on its broad shoulders, changes, of course, also the degree of density; and as water is naturally desirous to restore the equilibrium, there follows a constant flow to and fro; so that salt here appears as the great motive-power, which causes the currents of the sea! These again, in their turn, bestow warmth on Western Europe, mix the differently heated waters of the ocean so as to protect the life that teems in them against cold, and favor the sailing of trade-ships. Thus climate and temperature, winds and currents, navigation and the fertility of coast-lands, all depend on the presence of the little pinch of salt!

Far better known is the fact that man, like all animal life, cannot exist without salt, but must miserably perish, so that among the most terrible punishments, entailing certain death with fearful suffering, that of feeding criminals with saltless food was not uncommon in barbarous times, and prevailed, to our disgrace, until quite recently, in one of the northern countries of Europe. Animals, deprived of salt, lose their hair, become lean and hideous to look at, and die a death of unspeakable suffering. The reason is simple. A man, weighing a hundred and fifty pounds, carries in him at least one pound of salt; it constitutes five per cent. of the solid matter of his blood, and an almost equal proportion of all the cartilages of the body, and the bile contains soda as a special and indispensable element in the process of digestion. If the

salt, then, be withdrawn, or the ounce which every one
of us daily loses by perspiration and other means, be
not replaced, digestion is arrested, the bony part of our
frame is not rebuilt, the eye loses its brilliancy, and the
whole system breaks down.

Hence the craving of man and beast alike for the precious grain. Pliny but expressed the necessity of its
use for life, when he said that all the loveliness and joy-
ousness of life could not be better expressed than by the
name of salt, and the rulers of the world were not slow
in taking advantage of this fact, by taxing the indispensable gift of nature. Five hundred years before Christ,
already, the mythical king, Ancus Martius, established,
at the mouth of the Tiber, a saline, under the control of
the state; and at a later period the censor Livius earned
the name of Salinator, by raising the duty on salt.
From distant China to the west of Europe, every Government learned to treat salt as one of the regalia; and
not many years ago, poor French peasants were still
cruelly punished if they dared draw a bucket of water
from the great ocean, in order to secure the few grains
of salt it contained !

As vegetable food is both unpalatable and little nutritious unless accompanied by salt, herbivorous animals
everywhere delight in its use. The wild buffalo and the
deer, as well as our domestic cattle, enjoy it with evident
relish; and the Alpine herdsman, like the Gaucho of the
Pampas, trains his half-wild herds to meet him at certain

places, by depositing small quantities of salt at regular intervals. When the eager huntsman, in Southern Africa, is in search of rare sport, he hides himself at a favorite salt-lick, and is sure to be amply rewarded ; and the cunning chamois-hunter of the Alps prepares his way, years ahead, by cautiously placing a handful of salt in accessible spots, until even those sagacious animals are beguiled, by their greediness, and finally fall into the hands of their enemy.

Even here, however, man shows his strange superiority over lower beings; for while animals, without exception, love salt with equal fondness, the desire among men differs essentially. Nations who live largely on animal food, value it naturally less than those who prefer a vegetable diet. Thus Mungo Park speaks of certain tribes in Southwestern Africa, who never take salt by any chance, and adds that even Europeans, travelling in their country, never feel the want of it. The same disregard prevails in the colds of Siberia, where the peoples of whole districts eat their food without a particle of salt. On the other hand, there are Indian tribes, true vegetarians, who consume it in large quantities, so that the children are seen sucking pieces of salt like sugar. In certain portions of Africa, he is deemed a rich man who can afford eating salt with his food; in the mountains of the South, small pieces of it circulate as money, and on the Gold Coast a handful of salt will purchase two serviceable slaves.

A nicer distinction, yet, is the well-established fact, that the active races require salt more imperatively than the passive races; and this, in connection with the refined instincts of the body, explains, no doubt, the startling difference between the Gaucho of South America, who hardly knows what salt is, and the intelligent son of European races, who could not live a fortnight without his accustomed supply.

How wonderful, then, that the presence of "a pinch of salt," a thing of no value and hardly noticed by millions of us, should be the condition of animal and vegetable life on our earth! Truly, not only is man fearfully and wonderfully made, that his physical life and the activity of his heaven-born mind should depend on the little white crystal, but great are the works and wondrous is the wisdom of Him, who, from His throne on high, orders alike the heavenly bodies in their unmeasured space, and the invisible grain of salt in the bowels of the earth and the deep of the sea.

MINE OYSTER.

Oh sea, old sea, who yet knows half
Of thy wonders or thy pride ?

WHEN you visit the famous old town of La Rochelle,
with its Huguenot memories and its countless historic associations from the days of the great Louis to the
closing scene in the Napoleonic drama, you are most
likely invited to take a peep at the sea-farms, which are
the pride and the honor of that harbor. You push out
with rapid stroke or spread a picturesque but useful
little sail into "the sea, the open sea," you just begin to
feel the swell of the billows, and then you enter a rough
enclosure formed of huge blocks of stone, and are bid to
gaze into the depths, lighted up by a warm southern
sun, and to look at the living things innumerable which
there find a home in the mighty waters. There, near
the island of Rhé, you will be introduced to the new
sea-farms of our day, where not many years ago a row of
enormous and unproductive mud-banks stretched out
more than four leagues long, and where now, by a mira-

cle of enterprise and energy, some six thousand fishermen may be seen, as busy in their parks and claires as
market-gardeners in their strawberry-beds. You ask
what gives this multitude of men their lucrative occupation, and adds millions every year to the revenue of the
region around, and you learn with astonishment that it
is a scheme, first introduced by a stone-mason with the
curious name of Beef, to raise oysters !

If you have read your classics well, you may remember, at the mention of the dainty shell-fish, that there
was in Rome a man famous for the same bold under-
taking, who also bore a name of quaintest meaning.
This was Sergius Aurata, so called because of the number of gold rings he loved to wear, as some said, or,
according to others, because he was passionately fond
of gold-fish. He seems to have liked shell-fish even
better, however, for he was the first to transport oysters
from their birthplace on the coast to the Lucrine Lake,
where they were cleaned by the purer waters and fattened for the table, retaining their own native juices, as
Pliny tells us, and acquiring the flavor of their new
home. He must have been a pleasant man to deal with,
thanks probably to his intimacy with the delicate dish, for
Cicero sings not only the praises of his enormous wealth,
but calls him also a most pleasant and " delicious " person. To these attractive qualities he seems to have
added great cleverness, for he was at all times able to
supply the tables of Roman epicures with their favorite
8*

natives from his own park ; and so great was his renown
for ingenuity, that when he was sued in the courts and
threatened to have an injunction put upon his trade, his
advocate said defiantly, that if his client was prevented
from rearing oysters in the lake, he would grow them
upon the roof of his house.

They will, in all probability, present you with an oyster,
and ask you to taste its flavor.   Like all of us, you look
upon it simply as a delicacy, good to eat ; you open the
creature's rough and unsightly shell, and swallow the
delicate morsel to satisfy your craving appetite and to
please your palate.   But even the most refined and cultivated of oyster-eaters takes little note of the curious
intricacies of its organization, and knows nothing, nor
cares to know, of its wisely contrived network of nerves
and tiny blood-vessels.   In fact, men generally clip its
beard, that wondrous membrane of strange and curious
mechanism, by which the creature breathes, as thought-
lessly as they shave their own, and gulp down the luscious substance, unmindful that they are devouring a
body endowed with organs which all the science and genius of man has hardly yet been able to know and to admire, and which no power but that of the Most High
could ever devise and send forth into life.   They bolt
the living carcass, and decline being bothered and bored
in the act of cannibalism by the ill-timed and impertinent
interruptions of science.   And yet they are not the worst ;
for if Lucian already ridiculed the philosophers who spent

their lives inquiring into the souls of oysters, such wise-
acres were respectable, and the man who eats the oyster
with gratitude is at least excusable, when compared with
those who care neither for the oyster's soul nor its body,
but concentrate all their faculties on the shell. The sad
conchologist eviscerates the oyster as earnestly and as
gloatingly as the veriest Dando, but alas! he flings the
soft and savory substance from him, and delights in the
hard and unprofitable covering. His only pleasure is to
count all the little waves and scales and ribs, ill-shapen
and sad-colored as they seem to others, and he thinks not
of the living body within, as fearfully and wonderfully
made as his own.

Whilst, however, to the mass of men the oyster may
be nothing more than a rude and sportive device of
Nature, others, fortunately, have learnt to spell and to
read, to peruse and to study the great Bible of Nature,
in which this shell also is an humble letter, and they
have found out that the device is a sign pregnant with
suggestive meaning, carrying them onward and upward
to other forms higher in the scale of beings, and leading
them thus, with all things created, from Nature up to
Nature's God. But, to share in their joys and to receive
like rewards for our labor, we must first learn to approach all that was made with the reverence due to the
majesty of its Maker, and to be able to see half-hidden
grandeur in the minutest object, and veiled beauty in
the most ungainly creature. We must learn to estimate

each thing not carnally only, by its use and its pleasantness to our senses, but spiritually also, by the amount of
Divine thought which it reveals to our mind, believing
that every pebble holds a treasure, every bud a revelation. With such a spirit we shall soon find wonders in
every insect, sublimity in the tiny world of a pool, the
clearly-written records of past ages in a stone, and
boundless fertility of thought as of life upon the barren
sea-shore.

Even the life of a poor, silent shellfish, once reputed
the dullest and most inert of all animals, will then be
found to have its interest and its romance. In vain did
Plato already assign, in his transmigration of souls, people who, as men, were thoroughly ignorant and without
thought, to oysters thereafter, and speak elsewhere of
the soul being fettered to the body like an oyster to its
shell; in vain does Virey, in our time, call them the poor
and afflicted among the beings of creation, who seem to
solicit the pity of happier animals—they are, as we shall
see, beautifully made, capable of enjoying much happiness, and susceptible of being taught a lesson, which
most of us proud men have never been able to acquire.

Their life, usually pictured as one of utter helplessness
and unbroken seclusion, is by no means spent in unvarying repose. At the proper time, in the spring of the
year, when all Nature is full of tender love and restless
activity, the mother-oyster also is visited by the ruling
passion, and " the icy bosoms feel the secret fire." Soon

after, they are seen to contain a large quantity of milk-white fluid, which the microscope shows us to consist of almost invisible eggs and milt, lying snugly side by side in the same shell. Unlike most marine animals, however, the oyster does not heartlessly abandon its spawn and leave it to the mercy of winds and waves; but from the ovary the eggs pass into the sheltering folds of the mantle, where they remain for some time. Here they are surrounded by a nutritious substance, which serves to sustain them as the white of an egg supports the young chicken. After a while the whitish mass thickens, and oysters in this state are called "milky," because the mass of eggs resembles thick cream in consistency and color. The latter turns into yellow, then into darker brown, and the eggs are hatched! Suddenly the mother opens the shell; a dense mist is spread all around, and the young brood scatters far and wide.

Upon their first appearance in their new career, they are all life and motion, flitting about in the sea as gayly and lightly as the butterfly roams from flower to flower, or the swallow skims through the air. They are odd little cherubs, consisting, like the angels of old masters, of nothing but a couple of wing-like lobes on both sides of a mouth and shoulders, but not encumbered with a heavy, awkward body. The wings, fastened to rudimentary shells, are covered on the surface with countless little hairs, which move incessantly up and down, and thus enable the tiny creature to swim about in the water. Their infancy is one of perpetual joy and vivacity; they skip to and fro as if in mockery of their heavy and immovable parents. They do not go far from her, however, and the time of their joy is in their life, as in ours, but brief, and soon at an end. After a day or two they seem to have sown their wild oats, and if luck has favored them so as to escape the thousand voracious enemies that lie everywhere in wait or prowl about to prey upon their youth and want of experience, they finally settle down upon some suitable resting-place, a stone or a branch, and become steady, domestic oysters. But how few of them reach the goal! When they start from their mother's safe home, they count nearly a million; before they can find a new habitation, at least nine-tenths of their number have perished!

After they have attached themselves by means of a glutinous substance, with which provident Nature has endowed them, to some permanent place on what is called a good spatting-ground, the little wings, now useless, gradually dwindle and shrink, until they disappear, like the tail of a tadpole when it changes into the full-grown frog. Then they begin to grow, slowly, like all good things of this earth, from the size of a pin's head, at two weeks, to that of a pea, at three months; when they are a year old they are perhaps as large as a small lady's-watch, and at the age of five years they are in their prime. The shell remains frail and tender until they reach the size of that rare coin, an American dollar, but is hard and complete when they become fit for the table, which is in their fourth year. At that time they are rudely torn from their native bed by terrible iron prongs, to which they yield with philosophic resignation, and are carried unresisting to busy cities and the hum of crowds. If they should escape the gluttony of man, they die at the appointed time, leaving their shell, thickened by old age, and adorned with rings which show their years like the rings of a tree, to serve as a monument for times to come, and to add, with millions of their kind, a new layer to the crust of the earth.

Such is their life, simple and unromantic, but by no means as void of enjoyment as we are apt to imagine. There are countless sneers at the poor immovable oyster to be found in poet and prose writers, as if to be in perpetual motion was to be the perfection of happiness. The oyster has its time of merry wandering, when it is young; but it remembers, by times, that a rolling stone gathers no moss, and settles down quietly in its cool, pleasant home. We are so used to roam over the earth by rail and by steam, that we are apt to forget how Cain's curse was that he should be a fugitive and a vagabond all of his life! We learn a different lesson from the great Kant, whose philosophy DeQuincey praises above all ancient and modern wisdom, and who yet, never for a day, left his native town on the Baltic, and from thence wielded the lever that moved a world of minds; or from Burns, who said once that he envied only two beings in this world—a wild horse roaming freely over the steppes of Asia, and an oyster on a lonely rock in the ocean—the former had no wish it could not gratify, the latter knew no wish and no fear. Poor Burns preferred to lead the life of the horse, and we all know what came of it. Others have chosen the better part, and followed the example of the oyster, either withdrawing with stoic heroism into their shell, on which all the storms of fate could make no impression, or travelling sadly from Babylon to Jerusalem, from the wicked world and its tempest-tossed waves into the quiet convent, the peaceful haven on earth.

An oyster-bed in the sunny sea is the concentration of undisturbed happiness. The countless creatures congregated there may seem to be dormant, but we are sure they lead, each, the beatified existence of an epicurean god. The world without does not trouble them; its cares and joys, its storms and calms, its passions and sins, are all indifferent to the unheeding oyster. Apparently unobservant of what passes around, its whole soul is concentrated in itself, and like the sublime sage of the East, in his one word Om, the oyster finds bliss in simple existence. And yet it does not enjoy itself sluggishly or apathetically; its pleasures are neither few nor unvaried, for its body is throbbing with life and a thousand sources of enjoyment. The performance of every function with which the Creator has endowed them—and we know not yet half their number brings with it as much happiness as they are capable of enjoying.

The mighty ocean itself is subservient to their pleasure, and its rolling waves waft ever fresh and varied food within their reach. They have no care for the morning, for He who feeds the young lions provides an abundance for their wants; they need no effort, no labor, for the flow of the current brings the food to their very doors. Besides, each atom of water that comes in contact with their delicate, sensitive gills, sets free its imprisoned air to freshen and invigorate their pellucid blood. Nor can we doubt that the gentle agitation of the water as it flows around them, the equal temperature of the ocean, varying only from one degree of pleasantness to another, the act of imbibing the fluid and softly expelling again what is not required for breathing, that all these changes, unceasingly affecting their tender substance, afford them both wholesome occupation and cheerful amusement. We little suspect, when looking at the rough shell and the shapeless mass within, how beautiful the structure of the animal is, and at how many countless points it is susceptible to influences from the outer world. But if we put an oyster into a vivarium, and then aid our feeble sight by the inventions of science, we are struck at once by the millions of tiny hairs, cilia, which now are seen to vibrate incessantly, and to keep time most marvellously, as they beat on every fibre of each fringing leaflet. Even the very imperfect instrument in the hands of the great

Leeuwenhoeck made him exclaim with amazement : " The motion I saw was so incredibly great, that I could not be satisfied with the spectacle, and it is not in the mind of man to conceive all the motion which I beheld within the compass of a grain of sand ; " and yet his untrained eye saw but a tithe of what is now known to careful observers ! Well may we marvel, and adore the sublime goodness which devised all this elaborate and inimitable contrivance for the well-being of a despised shellfish.

As the oyster has a mouth, that also must be a source of enjoyment, although its suspicious nearness to the stomach deprives the mollusc, in all probability, of that enjoyment which the passage along the gullet affords to the gourmand among ourselves,—so that one of them wished it could be lengthened out into a mile. The oyster, however, has an appetite, and no doubt also its own power of appreciating the varied provisions with which it is continually supplied, and which are taken impartially from the animal as well as from the vegetable kingdom. It has its nervous system, moreover, very simple as far as we know, but connected with the ovarium, and thus affording the pleasant sensations of love ; the mantle, in whose folds its young are so tenderly kept for a long time ; and the heart itself, with its two chambers and its gentle pulsations, showing clearly that it feels and enjoys, though it may have but obscure sensibilities and limited instincts. Then there are still other portions of its frame, which the careless and the ignorant simply declare useless, because they cannot at once see what essential purpose of life they are made to serve, and because they might apparently be omitted without disturbing the course of daily duties. But as they are never found missing, and as we now know that nothing in created beings is the result of chance, we may safely assume that they are symbols of organs to be more fully developed in animals of higher perfection—anticipations, it may be, of limbs and senses given to other creations, and badges of the relationship which exists between these lower and despised beings, and man himself in all his sublime strength and beauty.

It is true, the oyster is not visibly endowed with other senses than taste and touch, which it exercises and enjoys in almost unceasing activity. We do not know that it ever ceases to take in food, and we can see distinctly that the beautiful cilia, more delicate than the costliest lace on the wedding-robe of an aristocratic beauty, shrink and shiver at the slightest warning, by day or by night. There is no outward eye perceptible, as, in fact, there is no head to which it might lend light and beauty in its dark home; and yet the oyster is exquisitely sensitive to every change of light, and finds in this susceptibility at least one means of protecting itself against an enemy. As soon as the shadow of a passing boat falls upon it from on high, and long before the pressure of the agitated waters can have reached its home on the rocks, it closes its shell, unfortunately with no better success than that of the cunning manœuvre of the ostrich, when it hides its head under a bush. The ear is, on the contrary, very fully developed, and a most curious organ, consisting mainly of a number of diminutive grains, shut up in a transparent prison, and there dancing in perpetual motion, which changes with every sound that strikes upon the outer walls. Here, then, is a new source of enjoyment, and the thousand subdued notes of the great ocean may have their melodies unknown to human ears, but appreciated by the dwellers in the vasty deep.

In spite of these organs, and the undoubted fact that oysters have senses and various sources of happiness, men have generally believed them to be very imperfect beings after all, and fit only to be mentioned among the lowest of created beings. But "there is a philosophy in shellfish, and above their jackets," in more senses than one, and whilst we have seen that they are endowed, in their own peculiar way, with sufficient acuteness and sensibility to make their so-called instinctive proceedings often very surprising, there are men, who know them well, claiming for them a certain degree of intelligence and thoughtful action. In fact, utterly helpless and thoughtless molluscs as they seem to be, they have proved themselves capable of learning that hardest lesson which man has to acquire in the world—to keep their mouth shut at the proper time! The manner in which they came first to be trained in this rare accomplishment was this:

There are large establishments on the coast of Calvados, like those near La Rochelle, where oysters are kept to be cleaned and fattened for the market. These artificial beds are constructed between tide-marks, and their denizens, accustomed to spend the greater part of the twenty-four hours under water, open their valves and allow the waves to come in, when so situated, but close them firmly when the receding tide leaves them exposed. Thus they get gradually used to these alternations of submersion and exposure, and the practice of opening and closing the shell becomes a regular habit. After a few years' residence here, they are ready to be carried to Paris; and as the distance is great, even by rail, this habit of gaping at a certain hour would insure their destruction, as the oyster can as little live without its supply of air, which it derives from the sea-water, as we ourselves. The French owners of these parks, therefore, undertook to train them to keep their valves shut in order to avert the evil. Each batch of oysters intended to make the journey to Paris, is now subjected to a preliminary exercise in keeping close even at such hours, at which the tide is in, by giving them at the right time a slight blow, which instinctively closes the door. The molluscs learn, after a while, to do so whenever they are uncovered by sea-water; and when the time for the journey arrives, they are tapped, and quietly close the shells, keep the gills moist with the water within, and arrive safely and lively in the great capital. Thus they prove themselves capable of understanding and profiting by a lesson, and are enabled to arrive in the metropolis like polished citizens of the Empire, and not like gaping rustics, with their mouths wide open.

The mollusc is, moreover, by no means so intensely selfish, that all the joys and pleasures connected with its existence should be strictly confined to its own secret life. In building up its house, for instance, it does not labor for itself alone. We cannot yet answer the question, which the fool asked of King Lear, how the oyster built its shell, but we can see with deep interest how varied its colors and how perfect its form. The upper part is generally raised—the oyster of Holstein alone has a concave top, having caved in, as the poor people say, when the Prussians took possession of the country—the lower part is flatter, only deep enough to hold some water, and both valves are movable by means of a powerful muscle, which holds the door more strongly than the best of our locks or latches. The outside varies according to the locality where the oyster grows: it is dark on black, muddy bottoms; the Spanish oyster is dressed in red, the Illyrian has a brown armor to protect its dingy body, the favorite of the Parisians is green without and within, and the natives of the Red Sea shine after the fashion of the Orient, in all the colors of the rainbow. At night the shell emits a dim, sulphurous light, arising from a variety of microscopic algæ, which enjoy their existence as parasites of the oyster. Nor are

the brilliant lustre and the gleaming iridescence of the
inner lining of the shell destined to remain hid forever in
the depths of the ocean. The nacreous shells, which furnish our mother-of-pearl, belong to a variety of oysters,
and are eagerly sought for, wherever they can be procured in sufficient quantities, forming an article of considerable importance in trade. Those of the tropics contain, however, still more precious treasures, for they
change the luckless grain of sand or unproductive egg
into costly pearls, and teach us the great lesson, that we
also should endeavor to treat our troubles in like manner, and convert our secret cankers, by help from on
high, into pearls of great value.

> " On some far-distant shores,
> There are who seek the oyster for the pearl
> She sometimes brings with her, a priceless dower—
> But Dando only sought her for herself."

And Dando was right; for what are all the beauties
of the shell, and all the charms of the rare pearl, to the
luscious food and the certain health promised to the
lover of oysters by the inside? Much has been said in
comic wonder and half-serious admiration of the man
who first ventured to eat an oyster. A quaint old German writer, Lentilius, said of the mollusc that it was " an
animal of horrid and nauseous appearance, whether you
look at it shut up in its shell or open, so that bold must
have been the man who first raised it to his lips." The
popular legend has it, that a man, walking one day by the

side of the sea " with its many voices," picked up one of these savory bivalves just as it was in the act of gaping. Observing the extreme smoothness of the sides within, he insinuated his finger to feel the shining surface, when suddenly they closed upon him with a sensation far less pleasant than he had expected. The prompt withdrawal of the finger was hardly a more natural movement than his bringing it to his mouth, by that unfailing instinct which comes to us in early childhood. In this instance the result was fortunate in the extreme. The happy owner of the injured finger tasted for the first time the delicious juice of an oyster, as the Chinaman in Elia's Essay, having burnt his finger, first tasted Cracklin. The savor was superb, and he had made a great discovery; he picked up the oyster, forced open the shell, banqueted upon the contents, and soon brought the mollusc into fashion—a fashion which, unlike all others, has never gone and never will go out again. To ascribe to the lucky man wonderful courage, is a vulgar error; he deserves admiration, on the contrary, for his highly sensitive and exquisite taste, and his prophetic appreciation of a dainty, as he saw the tempting morsel lie all succulent upon its own plate in its own delicious sauce. We can sympathize with the regret he must have felt, in common with all oyster-eaters, when gazing upon the entombed remains of millions of well-fed and elegantly shaped oysters, which geologists point out to us in the Eocene formation. We can imagine, with Mr. Forbes, how he would chase " a pearly tear " away, as he calls to mind how all these delicious beings came into the world and vanished to so little purpose.

Even when man and oyster were first brought in contact, they do not seem to have taken kindly to each other; at least, the silence of the Old Testament leads to the idea that to the Hebrews the shellfish was forbidden as one of the abominable beings. The ancient Greeks were far wiser in their generation, and enjoyed them heartily, but they deserve no mercy for the vile use they made of the outer shell. It was black ingratitude all around; for after having feasted upon the delicious oysters of their waters at the expense of some great patriot like Aristides, they escaped thanking him for his largess by writing his name on the shell, and banishing him from his native land. How could men blessed with luscious natives ever be guilty of ostracism?

The Romans, on the contrary, showed their appreciation of Nature's rich bounties by the fostering care with which they raised them, and the religious fastidiousness with which they prepared them for their enjoyment. We have seen already how they learnt to improve them, but they also took to importing them even from distant Britain, whose natives they prized above all others. Unfortunately, the gluttony of the time of the Cæsars affected their appreciation of oysters also, and a Vitellius could with beastly voracity set them the bad example of eating oysters at all times of the day, and up to a round

9

thousand at a sitting. To increase the heniousness of
the offence, he availed himself, in order to make this possible, of the abominable fashion prevailing in those days,
which made room for new supplies by removing the older
inmates through the agency of a peacock-feather, tickling
the palate with great effect. Seneca, who so admirably
'praised poverty in his writings, and complained on the
forum that he could not live comfortably with only ten
millions of dollars, treated oysters with the same du-
plicity. The temperate sage ate a few hundred every
day, until, in a fit of indigestion, and after having lis-
tened to a brother philosopher, who inveighed against all
the follies and vices of the times, he renounced them forever. With the bitterness of a friend changed into a foe,
he turned round and denounced them as vile things,
pleasing only to gluttons, because " they so very readily
slipped down and so very readily came up again." The
cooler Cicero, while confessing his fondness for them,
claims to find no difficulty in abstaining ; but the poets,
fortunately, speak with more enthusiasm. Horace, devoted to the delicious mollusc with his whole heart, sings
their praise again and again, and Juvenal breaks forth in
admiration of him

> " who could tell
> At the first bite, if his oysters fed
> On the Rutupian or the Lucrine bed,"

the Rutupians being the fine natives of Britain, which
had but just come into fashion. The Emperor Trajan

was so fond of them, that his famous cook, Apicius, had to provide them even during the summer months, and to send the master's favorite dish after him to Parthia, at a distance of many days' journey from salt-water. From that time onward, nearly all great men have been fond of oysters. Cervantes loved them, and satirized the oyster-dealers of his country; French authors professed a like fondness, from the learned doctors of the Sorbonne under Louis XI. down to the unhappy encyclopedists, who were joined by the great men of the Revolution in the days of their innocence. Nor have the Parisians degenerated since, for they still consume daily a million. Pope and Swift shared this partiality for oysters, and the Scottish philosophers of Hume's day spoke in raptures of their "whiskered pandores," an enthusiasm fully appreciated afterwards by Christopher North and the shepherd.

It is not the mere savor, moreover, which makes oysters such favorites among men, but they have valuable qualities besides, and have been recommended from of old by physicians of all countries for many diseases. It may not be true that their own fertility is transferred to those who eat them, as was fondly and firmly believed in former days; but there can be no doubt that they are marvellously nutritious, very digestible, and especially famous for their effect on the increased production of blood, so that they are often prescribed in cases of wounds or after repeated bleedings. Dr. Pasquier recommended them as curing gout, and Dr. Leroy, by taking two dozen every morning, preserved his youthful vigor to an advanced age. We need not wonder, then, that their consumption is enormous, and nothing can give a better idea of the quantity brought to market, than to see the fleet of oyster-vessels dredging in our great estuaries, or, what is perhaps even more impressive, to pay a visit to Billingsgate, the one great fish-market of the city of London. At the early hour between four and five in the morning, the visitor here sees one of the marvels of the overgrown city; the immense amount of fish of all kinds which London grasps by means of its gigantic iron arms, its railways and its steamers, from every sea that beats against the island-coast, and brings here in one point together. There he sees superb salmons, fresh from the friths and bays of Scotland, or from the fertile Irish seas, floundering about; delicate red mullet, all the way from Cornwall, which await being carried to the West End; smelts, with delicate skins varying in hue like an opal, brought from Holland in Dutch boats; pyramids of lobsters, a vast moving mass of spiteful claws and restless feelers, savage at being torn from their clear, cool homes in Norwegian waters; and perhaps a royal sturgeon of colossal dimensions, dragged with ropes through the excited crowd by a yelling knot of men. Among these there are heaped up such mountains of oysters as to appal the inexperienced, and down Oyster-street, as it is called, lie long

lines of oyster-boats, moored side by side, and heaping
full of natives and the lower kinds. And yet the railways bring in even larger supplies, especially since the
discovery of a great natural bed, called the Mid-Channel
Bed, which stretches for forty miles between the ports
of Shoreham and Havre, and has proved, as the dredging-
ground is free to all comers, a source of vast wealth.
Nor are private banks less remarkable for their extent;
so that long years ago a Mr. Alston, then the largest
oyster-fisher in the world, could, in a single year, send
fifty thousand bushels from one of his parks to London,
and pay eight hundred pounds metage to the owners of
the market. The whole supply, now, is stated at eight
hundred millions a-year, and yet there is a pause, at least
during a part of

> "those four sad months, wherein is mute
> That one mysterious letter, that has power
> To call the oyster from the vasty deep."

The question has often been raised, why, if oysters are
really the greatest of gastronomic blessings, and life is
proverbially short, the dainty creature should not be eaten
all the year round. The prejudice, however, which for-
bids them during the months that have no letter R in
their names, is not altogether unfounded. In May and
June they generally spawn, and then their life's blood is
essentially changed for the benefit of their posterity, and
their own flesh is lean and unpalatable. Besides, however productive they may be, a conscientious lover of the

mollusc will hardly reconcile himself to the barbarous
waste of swallowing with each living parent a million of
promising offspring.   In the next two months the heat is
apt to be so great as seriously to endanger all oysters
that are not eaten immediately after they are taken from
the water; and one spoiled oyster does more harm than
a thousand good ones.   Hence the English rarely have
them brought to market before the first days of August,
when the " common oysters " from Colchester and Fever-
sham appear gradually, but the "melting natives" are
not seen before the beginning of October, reach their
meridian of perfection at Christmas, and disappear again
towards the end of April.

In the remaining months, however, they throng the
markets of the world, and then they are eaten by old and
young, by rich and poor, " the only meat which men eat
alive and yet account it not cruelty," as old Fuller says
quaintly.   For this is their great merit, that one may eat
them to-day, to-morrow, and forever, and as many as one
wants, and yet their presence hardly makes itself felt,
while they gratify the palate, quiet the excitement of
certain nerves which we call hunger, and leave no feeling
of satiety, no reproach, no remorse for the following day.
They are the true *grata ingluvies* of Horace.   Hence we
marvel how a clever man like Malherbe could say that
he knew nothing nobler in the world than women and
melons, and yet, living as he did on the coast of Normandy, and near the finest of oyster-banks, forget oys-

ters ! We all know men with whom women do not agree, and how many of us can eat melons with impunity; but who ever heard of fresh oysters making themselves at all disagreeable ? They can, moreover, be eaten at all times of the day; they are good at breakfast, excellent as a prelude to dinner, and Juvenal speaks already of his beloved Venus Ebria,

> "Who at deep midnight on fat oysters sups,
> And froths with unguents her Falernian cups."

The true way to eat them, profitably to taste, health, and enjoyment, is, of course, to eat them raw, and without condiment ; for vinegar, pepper, or lemon-juice, all spoil the natural flavor of the bivalve. The only good dressing is its own gravy, which is not sea-water, as many fancy, but its life's blood, which it sheds when the shell is violently broken open. Hence a master of the art says of all other ways of dressing : " Frivolity ! profanity ! sacrilege ! If after such treatment they taste well, they are no longer oysters ; if they are still oysters, they have no longer any taste; " and the poet adds sagely, that in his view oysters ought to be eaten, as we love to see white roses,—with the dew of a fine summer morning on their tender leaves. To all of which famous Dr. Kitchener adds, with refined cruelty : " Those who wish to enjoy this delicious restorative in its utmost perfection, must eat it the moment it is opened, with its own gravy in the under shell ; ·if not eaten absolutely alive, its flavor and spirit are lost. The true lover of an oyster will have some regard for the feelings of his little favorite, and contrive to detach the fish from the shell so dexterously that he is hardly conscious he has been ejected from his lodging till he feels the teeth of the gourmet tickling him to death." Would Dr. Kitchener be very grateful for being tickled to death?

If dressings are not allowed, some drink to accompany the mollusc on its way is generally considered indispensable. Strong wines and liquors should be eschewed, although in this country whiskey or gin, and in Germany and Russia rum, is taken with them; these beverages simply pickle the oyster at once, and deprive it of its best qualities as nutritious, digestible food. Lighter French wines are less objectionable, such as Chablis, Sauterne, and even Moselle, but Port is said to turn them into stone; porter and ale, on the contrary, and better still half-and-half, are considered the true friends of the oyster.

The question as to how many may be eaten at a time is fraught with great difficulty, for here men differ as well as doctors. The experienced say that oysters after the fifth or sixth dozen cease to be a delight; specially favored individuals speak of seven or eight as profitable in times of great political or domestic excitement, when the system has to be appeased by a specially cooling and soothing food. But Brillat Savarin, in his admirable book on Taste, expressed a different opinion. " It is well

known," he says, "that formerly, under the Louises, before the Revolution, every festive meal began with oysters, and that a certain number of guests were always
found who did not rest until they had eaten a gross, viz.,
twelve dozen. The abbés of those happy days, especially, never were content with less, and the chevaliers
often went beyond them. As I wished to know the exact value and weight of such a preparation for a good
meal, I took my scales, and found that twelve dozen
oysters, with the water they contained, weighed exactly
three pounds. How much happier, now, were these
worthy guests with such a weight of oysters than if they
had eaten three pounds of meat, or even of poultry!" A
handsome compliment, surely, to our friends the oysters,
which could not have been more happily turned by—the
best of cooks. In another place he adds a remarkable
instance of individual capacity. It seems that he accidentally fell in, in 1798, with a certain Laperte, an officer
in one of the public courts, who professed to be passionately fond of oysters, but never to have had, as he said,
" his fill " of them. The author offered to give him that
satisfaction, and invited him to dine the next day at his
house. The gourmet came, and Brillat kept him company up to the third dozen, when he let him go his
way unaided. He marched on bravely, till he reached
the thirty-second dozen, which he did in about an hour,
as the man who opened the oysters was not very expert.
Brillat became impatient, not at the endless capacity,

but at his own forced inactivity, thinking it both "painful and unwholesome to sit at table without eating," and stopped his valiant guest in the midst of his exploit. He expressed his regrets that the Fates had evidently denied him the privilege to let his friend have his fill that day, and invited him now to join him at dinner. The guest assented, and behold! to the author's amazement, he went to work with all the energy and perseverance of a man who had sat down to table after long fasting!

It is not impossible that this happy Laperte may have belonged to the school of the poet Lainez, in Paris, who was asked, after four hours' active devotion to an uninterrupted dinner, if he had dined yet, and replied, indignantly: "Do you imagine my stomach is endowed with memory?" Whereupon he resumed his work with renewed zeal and increased vigor.

There is comfort in the thought that even in such extreme cases no man has yet been known to have suffered seriously because he loved oysters "not wisely but too well." There is comfort, also, in the fact that all the voracity of man could make no impression on the vast numbers of oysters which exist in our seas. Spenser already said, it was

"much more eath to tell the stars on high,
Albe they endless seem in estimation,
Than to recount the sea's posterity;
So fertile be the floods in generations,
So huge their numbers, and so numberless their nations."

Natural beds and banks of oysters are found in all the

seas of the temperate and torrid zones, now stretching out miles after miles in all directions, and now rising so high that ships are wrecked on their crests. And thus it has been apparently from time immemorial, for gigantic structures, consisting of fossil oysters, are found in many places. In Berkshire, England, a petrified colony of oysters covers more than six acres ; in Massachusetts and Georgia enormous breakwaters are formed between the firm land and the hungry ocean, ramparts twelve to fifteen feet high, the lower layers of course fossil, but the upper strata alive, and affording delicious food to the negro of our day, as their forefathers did to the Indians, and perhaps to the Aztecs. On the west coast of this continent vast surfaces are covered with fossil oysters, which have been raised by volcanic action, and now tower to the height of sixty feet and more, for thirty miles at a time.

Among the living, however, there is as great a difference as among the races of men. Those of our country are acknowledged to surpass in size and luscious flavor all others; and even English travellers, like Charles Mackay, have acknowledged them to be superior to the famous Whitstables at home. But Frenchmen, accustomed to their own smaller and richer oysters, with a strong taste of copper, object to their inconvenient dimensions, and miss the metallic flavor. Germans utterly at sea in all that concerns the sea, either do not appreciate oysters at all, or, if they do, are enraptured by the ample provision contained in each shell and the amount of lager it requires for easy conveyance. Next to our own come undoubtedly the English oysters, of which there are many varieties, the best growing on submarine rocks, an inferior kind on sandbanks, and the coarsest on muddy bottoms. England values them largely according to size, and sends the smallest kind, called Dutch-size, over to Holland. The common oyster from the Western coast is very large, with thick shells and little meat. The Colchesters go by the name of Middle Ware, and are larger than the best kind, the Little Natives, reared carefully at the mouths of a number of small rivers and in Southampton Water. Scotland is justly proud of her Pandores, so called because they are found near the salt-pans in the neighborhood of historic Prestonpans, and caught, it is said, by a bit of magic. The fishing-crews keep up, while the dredging is going on, a kind of wild monotonous chant, to which they ascribe great virtue, and sing:

> " The herring loves the merry moonlight,
> The mackerel loves the wind ;
> But the oyster loves the dredger's song,
> For he comes of a gentler kind."

Paddy claims for his Pooldoodies of Burra, and especially for his Carlingfords, that they are superior to all the world, and is as usually correct in his patriotism, but mistaken in his assertion. They are very fine, however, with a dark, almost black beard and delicious flavor, but

not to be compared to some of our own varieties. The
natives of England are largely sent over to Ostend to
be cleaned and fattened in Belgian parks, and then
assume a perfection almost unsurpassed. The shell
becomes very fine, almost transparent; the fish is small,
but rich and beautifully white, and bearing to the best
of common oysters the relation that a well-fed capon
bears to an ordinary chicken. This is the oyster which
gourmets prefer to all others. It goes from Ostend all
over Germany, to Russia, and even to distant Odessa.

French oysters are limited to northern seas, the Mediterranean coast having none that are worth eating. Those
raised at Marennes, in the Bay of Biscay, and at the
Roches de Cancale, are the most famous, though the whole
coast, from Normandy to Dunkirk, abounds in excellent
kinds; they are brought, to the amount of about two
hundred millions a year, to the Rue Montorgueil, which is
to Paris what Billingsgate is to London. The most
striking feature, however, is the preference which Parisians give to green oysters, and the pains which are therefore taken to produce the color artificially, by favoring
the growth of certain sea-algæ. These parasitic plants,
when once introduced into oyster parks, soon cover the
walls and rocks, and gradually spread their transparent
veil over the molluscs themselves. The adversaries—for,
like all superior things in this world, oysters, and especially green oysters, meet with opposition at times—say
that the green matter enters into the gills of the luckless

creature, stops the breathing, and thus causes dropsy. The disease makes the oyster to swell, by which process the texture of its meat becomes looser, finer, and more palatable ; and epicureans revel in dropsical shellfish as they delight in diseased goose-livers. The Baltic has a small supply of the precious molluscs, but the variety is coarse and insipid, probably because the waters of that sea are not salt enough ; those of the Adriatic, however, and of the Bosphorus, are better, and in great demand during the long fasts of the Greek church.

Wherever the oyster, therefore, appears in sufficient quantities, there men are found ready to consume them as fast as they can be procured; but the poor unselfish oyster has enemies nearer home, in its own native element, and close upon its borders. The arch-enemy is the sleepy, stupid-looking starfish, the Master Five-fingers of our boys, who eats them as spat, or even when grown to considerable size. These greedy devourers have the curious power of rolling themselves up and floating away, so that they appear and vanish again, no one knows how. But all of a sudden, and often at the very time when the sanguine fisherman ·gets ready to reap a rich harvest from a well-stocked oyster-bank, he finds, upon coming to the grounds, that the foe has been there before him, and millions of starfishes have settled down like a flock of wild pigeons on a field of wheat. Generally, they prefer the spat or very young oysters, which they take whole into their capacious mouths, and there digest slowly. But how do these tender, fragile creatures manage to get at the full-grown mollusc in its impregnable fortress? The ancients had a story, that they watched it till they found it incautiously yawning, and then slyly slipped their greedy fingers between the valves to keep them open, while they devoured the contents. This is, of course, a mere fable, as the soft, slimy finger would be squeezed off in an instant, even if the starfish were not famous for falling to pieces by immediate suicide as soon as it is brought into contact with a hard substance. Its murderous assault is far more curious. The first step in the process is for the enemy to lie close upon its prey, folding its slimy arms tightly over it, so as to hold itself in the right position. Then it applies its mouth closely to the victim, and as it cannot, by any force of its own, put the oyster into its stomach, it deliberately proceeds to put its stomach into the oyster! It begins slowly but steadily to push out this organ through the mouth, and wraps the mollusc in the folds of that capacious bag; patience always does its work, and in due time the hapless native surrenders to the devourer.

Another enemy shows, if less originality, at least equal perseverance. This is the whelk, who also seems, like the vulture, to smell its prey from afar, and although endowed with very slender means of locomotion, appears in vast multitudes, when least expected, on the oyster-beds which it deems ready for use. It assails the shell boldly from above, and with marvellous patience drills, by means of its sharp tongue, a hole in the upper valve, by which it gets at last fairly inside, and then enjoys the dainty food. Mussels come by myriads, when young, and cover the luckless oyster with a fine, ropy texture, which catches mud and sand, and finally smothers them; and gray mullets appear in swarms, and, greedily grubbing, devour whole beds of well-fattened natives. Even the elements combine against the helpless mollusc; heavy gales of wind at times roll them up in ridges three feet deep, when mud and seaweeds settle on them and choke them speedily; or frost and snow and ice kill large numbers, when they are not safely sheltered at a depth of at least three or four feet of water. Thus it is, that by the wise provisions of Nature, the danger of overstocking her vast reserves is avoided; for wherever animals multiply their species at such enormous rates, there are, on the other side, numerous enemies ever present to keep it down and to prevent an undue preponderance.

All the voracity of man, however, and all the persecution of enemies, does not destroy enough oysters annually to prevent them from forming, as we have seen, gigantic deposits in various parts of the globe. For, if left to themselves, oysters grow old and die a natural death, though it has not yet been ascertained fully what age they are allowed to reach in their solitude. The expert fisherman, it is true, can tell at a glance and to a nicety the precise age of his flock. He examines the successive layers on the upper shell, technically called shoots, and as each of them, overlapping the lower, marks a year, he is at no loss to ascertain how old the house and the inhabitant—for they are always of the same age. These layers, it seems, are regular, and laid in even succession one upon the other, until the oyster attains its maturity, which is generally fixed at seven or eight years; but after that time they become irregular, are recklessly piled upon each other, and make the shell look bulky and ill-shapen. As some molluscs have been found with shells nine inches thick and of a perfectly enormous size, it is fair to presume that the oyster, when left to its natural changes and unmolested, may reach a patriarchal age, and even outlive our race.

Unfortunately, man nowadays rarely allows them to pursue the even tenor of their life. On the pretext of protecting them against their powerful enemies and of improving their race—pleas not quite unknown to certain nations of our day—they are taken when quite young from their home, and brought to so-called sea-farms, where they live, safe against all danger, well fed and happy, and reward the favor shown them by increasing at least to double their value. Little is known of the labor and expense, the care and attention bestowed upon the apparently trifling mollusc, in order to make it acceptable to fastidious palates or even simply fit for market. First, the spat, or fecundated sperm, is stored up in large vats, specially designed for the purpose, and thus the immense quantity of seed-oysters are saved, which on natural banks fall an easy prey to countless devourers. These are sold as Native Brood to dealers in the article, and conveyed to artificial ponds or reservoirs, called oyster-parks. These receptacles, which are often of vast size, have a floor of clean stone slabs, covered with fine sand, on which the small oysters are carefully laid on the proper side, and a little inclined. The sea-water is made to enter gently, so as not to wash sand into the shells, which would kill them instantly, and rises and falls with the tide outside. If the oysters are to be very large and of light color, each tide must bring fresh water; but if they are to be delicate and of finer taste, the water is allowed to remain some time in the basins, so as to favor the development of the microscopic plants, which are always present in sea-water and largely form the food of the oyster. Here they are kept generally three or four years, till they have reached a good size and are considered fit for consumption. So far, their education has been left largely to Nature; but now additional steps are taken to perfect their condition, if they are to bring specially high prices. They are stored in large, shallow vats, where they gradually get rid of the taste of mud, which many still have, especially when they come from beds and banks situated at the mouths of rivers. Here they are simply kept in fresh sea-water; the method of fattening them with oatmeal having been given up, as the throwing in of dead stuff only makes the water foul and the oyster sick, and because very fat oysters are considered, like prize cattle, none the better for overfeeding. Such oyster-farms exist now in large numbers, mainly in England, where a single private oyster-park, near Whitstable, is valued at two millions of dollars; and in France, where the Government, true to its fostering policy, supports the enterprises by every means in its power.

When the poor oyster leaves these vats, it approaches its tragic end, which it reaches only after much tribulation. The journey to the landing-place, whether it be a pier in the river or a railway-station, is generally pleasant enough; they are transported carefully, travel in good company, and are occasionally refreshed by supplies of new sea-water. But when they arrive, the bad treatment begins; they are pushed into baskets, tossed into barrels, pitched on carts, fortunate, yet, if a kind hand brings them at intervals a pittance of water. Too often, however, the same hand gives them a stone instead of a loaf, for the common error still prevails, that salt and common well-water will do as well—a cruel mistake, since it is neither the salt nor the water which sustains the life of an oyster, but the abundance of invisible plant-seeds and microscopic spores contained in sea-water, which kitchen-salt kills on the spot. At last they reach their goal: if handsome, well-shaped, and well-flavored, they are introduced to the palaces of the rich and the

noble, to give, like wits and poets, additional relish to
their sumptuous feasts; but if sturdy, thick-backed,
strong-tasting creatures, Fate consigns them to the capacious tubs of common carters; they are dosed with coarse
black pepper and pungent vinegar, and depart this life,
partly embalmed after the manner of ancient Pharaohs.

LIGHT AT SEA.

" The swift-winged arrows of light."—ALEXANDER SELKIRK.

FAR out in the great harbor of Portsmouth, at a distance of over fourteen miles from the town, there
rises amid the stormy channel waves, a group of rocks,
the dread of all sailors, and the scene of dire disasters.
For more than two centuries efforts have been made to
place a beacon there, to warn the vessels which continually throng the great thoroughfare, and to prevent
the further loss of precious lives, and not less than ten
structures have, one after the other, been erected on the
ill-fated spot. It was only after many sad failures and
bitter disappointments that at last the true principles
were discovered upon which light-houses ought to be built,
and the noble tower of Eddystone could be raised, one
of the greatest glories of England.

The first of these experiments was perhaps the drollest
ever made, and worthy of its eccentric author. E. Wistanley was already well known all over England as a

man of wonderful learning and unbounded resources of mind. His home in Exeter was looked at with fear by his neighbors, and had a strange fame abroad. Visitors, whom curiosity or friendship had drawn there, came back with strange accounts of slippers, which had no sooner been touched by their feet, than ghosts, gaunt and grim, had arisen before them; of easy chairs whose arms had seized them as they sat down, and held them with an iron gripe; and of charming bowers in the trim garden, which greeted the thoughtless wanderer with showers of water. How he came to think of building his first light-house, is not known; but he certainly deserves all the more credit for his humane enterprise, as he lived in an age (1696) when old women were still thought capable of raising a storm in the channel by their incantations, and the church thundered anathemas against all who dared interfere with God's own appointed messengers—the winds. The structure which he erected at his own expense was as odd as the times and the character of the builder. It had open galleries running around the tower, and was adorned in all its stories with enormous beams, projecting far over the rocks, and intended to serve for cranes and pullies. It resembled more a Chinese pagoda, ornamented with quaint devices and gigantic inscriptions, than a light-house of our day; and not the least ludicrous feature was a covered balcony in one of the upper stories, from which the benevolent builder loved to fish with a rod. He used, in the pride of his success, to step out on this balcony and boldly defy the storm, crying aloud : " Blow, oh winds ! Rise, oh ocean ! Break forth, ye elements, and try my work ! " The winds were not long in coming, nor the ocean in rising : on the 26th of November, 1703, Wistanley had gone to his tower in order to superintend some repairs ; during the night a fearful storm arose, and on the morrow the sea had swallowed up the tower and its author.

It is true that this was one of the most fearful storms ever known in the Channel ; thirteen men-of-war perished in it, with 1,519 souls on board, and among them the " Mary," commanded by Admiral Beaumont, which disappeared entirely on the Goodwin Sands. This immense loss of life, the injury inflicted on the trade—Bristol lost 150,000 pounds, and London a million—and the terror it had inspired on land even, made so deep an impression that a public fast was ordered by the authorities to avert the wrath of God by deep penitence. By one of those curious coincidences, which no doubt occur frequently but are noticed only in rare cases, the model of the famous light-house in the library of Wistanley's house at Littlebury, Essex, was thrown down and broken at the same moment when the structure itself was blown down by the storm, at a distance of over 200 miles.

The second effort was made by a remarkable man, Rudgard, a silk merchant of London, who had a natural talent for engineering, and who, from great grief for the repeated losses and a laudable ambition to become a public benefactor, resolved to give his time and his money to the erection of a new light-house. With the aid of only two carpenters and three workmen, he set out on his labor of love, but, though as ill prepared as his unfortunate predecessor, his plan was far superior to Wistanley's, and possessed certain marks of such undoubted genius, that Smeaton, the highest authority on such subjects, speaks of it with unbounded admiration. Instead of giving his structure a number of corners and openings, by which wind and waves might attack it, as his predecessor had done, Rudgard built a simple, solid cone, well fastened to the foundation, and presenting no hold to the elements. It might have stood to our day, if it had not been built in alternate courses of wood and stone, for it had successfully resisted the storms of forty-six years, when it fell before a more dangerous element even than air or water. On the 1st of November, 1755, when one of the watchmen went up to the highest story to snuff the candles, he found the lantern on fire. He knew not how it had broken out, tried to extinguish it, and finding his efforts unavailing, called upon his two comrades to help him. They did not hear him at once, and, faithful to his trust, he remained at his post. In the meantime, the roof had begun to melt, and a shower of molten lead fell upon his head, his shoulders, and even in his mouth; he was at last carried out, but died on the twelfth day after the occurrence, and the physician actually found a small piece of lead in his stomach!

His companions had fortunately been able to escape to a neighboring ledge of rocks, from which they were rescued on the next day by fishermen whom the bright light had attracted.

The light-house at Eddystone was farmed out like all others, and this made it the interest of the so-called owners to rebuild it quickly, as in the meantime no dues could be collected from passing vessels. The absence of the light, moreover, caused at once several disasters, and public indignation rose to a high pitch. It was then that the great Smeaton was called in, and gave it at once as his opinion that the new light-house should be built of granite. At first they were disheartened by the time and the large capital such a work would require, but they prudently yielded to the suggestions of the great engineer, and have been well rewarded for their wisdom, for his noble structure is still standing, and proudly claims the first rank among the light-houses of all nations.

Little as we may credit the stories of Newton's apple falling from a branch overhead, or of Watt's tea-kettle sending forth clouds of hot steam, there is no denying that Nature is ever ready to furnish us with suggestions, and that the careful observer has no better teacher than her. Thus it was with Smeaton, when he was trying to solve the great problem how such a tower as he proposed could be made fast to the rocky foundation. He had been in the habit of walking over the flat country near

10

Plymouth, and of resting at times under a clump of trees to enjoy their grateful shade. On one occasion he found that they had been torn up by the roots during a fearful storm; one only of the whole group had braved the tempest, and stood unharmed in all the pride of its strength. Smeaton examined the oak which had shown such marvellous power of resistance, and he found the solution of his problem. The light-house was rooted to the rocks like a tree, holding the ground all around it in its powerful grasp.

The first stone of the great monument—for such it really is—was laid on the 15th of June, 1757, and the last was added on the 24th of August, 1759. Arising boldly from the bare rocks, famous for their terrible eddy, standing free and fearless among a mass of white foam that continually dashes against its base, it presents the appearance of a single solid shaft. It looks like a monolith, so accurately and carefully are all the blocks joined, dovetailed, and cemented; and when the waves rise and the sea washes over it all, dashing its foamy crests high above the lantern, it strikes the beholder with amazement and wonder. Very different from the boastful inscriptions of Wistanley are the words carved on Smeaton's work: " Except the Lord build the house, they labor in vain that build it," is written on the lowest course, and the keystone above the lantern bears the simple words: *Laus Deo!* expressive of his joy and his thankfulness. While he was yet at work, a party of French Corsairs, availing

themselves of the war then going on between England
and France, landed near the rocks, seized the workmen
and threw them into a French prison. Fortunately Louis
XIV. heard of it, and immediately ordered their release,
saying : " I wage war against England, but not against
mankind ! "

If England has thus avowedly rendered a signal service
to all nations, and is justly proud of her Smeaton, through
whose genius and great skill this was accomplished, the
sister kingdom can boast of her Stevenson and his great
work on Bell Rock. About twenty miles south-west of
Red Head, in Forfarshire, there lies a reef of rocks in the
German ocean which has cost countless lives, and is a
perfect grave-yard of vessels. The reef is about 2,000
yards in length, and although the surface is uncovered
at low spring-tides, the depth of the sea at only a hundred yards' distance from the rock exceeds three fathoms.
For ages seals and cormorants had been the only inhabitants, and yet so dangerous was this treacherous reef
that centuries ago the good monks of Arbroath, settled
there by William the Lion, placed a huge bell on the
rock, which rang by means of machinery set in motion
by the waves themselves. Hence the name of Bell Rock.
It was not until 1800 that Robert Stevenson, then as yet
little known, was employed to survey the reef and to report on the feasibility of erecting a light-house on the
spot. The main difficulty arose from the fact that the
reef was covered at high tide with several feet of water,

and the absence of any shelter for the workmen on the rocks, which lay isolated amid a waste of waters and far from any human habitation. To provide a home, a patache—a small vessel of peculiar construction—was anchored near the rock and provided with a beacon, while another vessel was engaged to keep open the communication with the shore. Workshops, in which the granite for the outer coat and the sandstone intended for the inner lining were to be prepared, were built at Arbroath, and a large dwelling-house was rented for the laborers; the latter, however, were expressly bound to remain a month at a time on the reef. On the 7th of August, 1807, Stevenson and his assistant, Peter Logan, with a few workmen, proceeded to Bell Rock, chose the precise spot where the light-house was to stand, and immediately went to work, first of all cleaning the place of seaweeds and incrustations.

The work was one of extreme difficulty, as the men could only labor in the intervals between the tides, and had to run for their lives every two or three hours. The wind and the weather furnished other interruptions, and although the men were found willing to work on Sundays, and even at night by torch-light, the structure made naturally but very slow progress. Accidents happened and great dangers were incurred. Thus once the sloop "Smeaton," which was used as patache, was torn from its moorings and carried with it a small vessel full of workmen; the tide was just receding, and to return was impossible till the reef should be once more completely submerged. There were still thirty-two men on the rock, and the two remaining vessels could not have taken in more than half of them if the sea had suddenly risen. Fortunately they were ignorant of their danger, as they were almost all sitting or kneeling at their work. Smeaton and one other man alone were cognizant of their great peril. But when the rising tide drove them from their work and they ran for the vessels, their astonishment was great at finding only two instead of three. Still, as Stevenson himself tells us, not one uttered a word; they only looked at each other, counting their number in silence, and showing in their features alone the uneasiness they felt. At the critical moment, however, a boat arrived from Arbroath with the mail, by the aid of which all were embarked and went in search of the run-away patache.

Another accident was even more formidable by occurring at night. The wind rose very suddenly and a heavy swell washed over the rock; the boats were instantly stranded on the wharf and threatened every moment to be upset. The torches went out and the darkness of the night appeared in all its horror. The sea was overcharged with electricity, and as the waves were dashed into foam against the rocks, liquid flames seemed to cover the reef. This gave to the ocean a terrible majesty, and the bravest trembled for a moment. Many features of this Crusoe-like life were of great interest to an intelligent observer like Stevenson. He was struck with the elasticity of the human mind, when he saw the wearied workmen, as soon as the tide put a stop to their work, hasten to their varied amusements. Some loved to read, and seized their books with eagerness; others preferred music, and the violin or the flute enlivened the strange scene; still others preferred fishing. Their only enemy was sea-sickness, which attacked them all, and hardly left them even in the course of time. Nor was it less surprising to see how soon these landsmen learnt to jump into the boats and to creep again out of their narrow berths, or to watch them like the Israelites of old, work with a tool in one hand and a lighted torch in the other, undisturbed by the roaring of the tempest and the booming of the waves. During the winter the work ceased altogether, but it was a great satisfaction to the builder and his men, that on resuming their labors in the succeeding spring they found the building perfectly sound and uninjured by the terrific storms of the winter. When the last stone of the twenty-second course had been laid, prayers were held for the first time in the new tower; all the workmen were voluntarily present, and two of them joined their hands to form a pulpit, on which the Bible lay during the service. At the end of 1810 the authorities took possession of the completed structure, and on February 11, 1811, it was lighted up for the first time. Those lights have never yet been allowed to disappear for a night, and the house has required no repairs for these fifty-seven years.

There is something touching in the manner in which the son of the great engineer, Alan Stevenson, repeated a generation later the work of his father, and built upon Skerryvore a similar structure in the open sea. He modestly acknowledges that whatever superiority there may be in his work is due to the lessons learnt from his father's experience, and the effective aid of the new agent, steam, which was unknown in the elder Stevenson's time. It is due to these three men, Smeaton and Robert and Alan Stevenson, to state that they were the first who dared build such massive structures surrounded by the waves, and well may we imagine the ocean saying to them in the poet's words: .

Great I must call them, for they conquered me.

Little that is really new and important has since been added to the art of building light-houses, if we except the truly American enterprise of carrying a whole building of the kind from one place to the other, as we see houses in our cities leaving their old foundations, wandering through the streets and seeking afar off a new home. This occurred in Sunderland, where the repairs in the harbor led to the construction of a new jetty, rendering the old one useless, and involving the necessity of building a new light-house on the new wharf. Workmen had already prepared the work of demolition, when a clever engineer,

John Murray, conceived the happy idea of conveying the whole building as it stood to the place on which the new tower was to be erected. The enterprise was far from trifling. The distance was 475 feet; the line of transportation a broken one, so that the tower had to be entirely turned round; the old wharf was a foot and seven inches higher than the new one; and what seemed to be the most serious obstacle, the octagonal towers, built of Portland stone, and 76 feet high, rested on a base of only fifteen feet. But the engineer provided for all these difficulties. Openings were made in the foundation course, through which powerful oak-joists were passed from side to side; these were bound together by equally strong timber in the opposite direction, and when the small portion of the base below had been knocked away, the tower virtually rested on a powerful platform, fully capable of bearing its weight. Next, strong stays, joined by stout cross-pieces, were built up all around the light-house, surrounding it with a timber-cage. One hundred and forty-four heavy cast-iron wheels were placed under the whole, to run on eight iron rails, which had been laid on the brickwork of the wharf and the jetty. As the colossus proceeded, these rails were taken up behind and laid down again in front, while a number of workmen were placed at the windlasses, which pulled the platform with its enormous burden by means of iron chains. The whole operation was so well prepared that it only required a little over thirteen hours to move the immense structure into its new position, and on the same evening the light shone once more as it had done the night before, over the harbor and the dark waters beyond.

The finest of all light-houses on the coasts of France is strangely enough also the oldest of the whole family. The Tower of Cordonan crowns a rock at the extreme mouth of the Gironde, and shines upon the stormiest part of the ill-famed Bay of Biscay. Far more turbulent than the British Channel at the most dangerous point; far more treacherous even than the waters that struggle forever against the basalt rocks of Brittany; the Bay of Biscay, in the line from Cordonan to Biarritz, is the most pitiless of them all. Turning suddenly to the southward, the sea seems to encounter an abyss below, into which the waves rush furiously, only to escape from the terrible pressure and to rise to a greater height than is seen anywhere else. Hence, from time immemorial, efforts have been made to diminish the dangers which lurk under these waters, and to furnish a safe entrance to the countless vessels that seek the large river. A ledge of level rocks fortunately offered from olden times a secure lodgement, although it has always been covered and washed by the tides. As early as the thirteenth century, merchants who came from foreign parts in search of the warm wines of Bordeaux, urged the importance of lighting up the unsafe harbor entrance. It was long believed that the great city of Cordova in Spain was especially interested in this matter, as her wealthy traders carried to the Gironde the

skins and the leather for which the city was renowned,
and that hence the locality itself obtained the name of
Cordonan.   It is certain that the Black Prince ordered,
about 1364, a light-house to be erected there.   It was, of
course, of the simplest kind—a mere wooden structure of
46 feet height, ending in a platform, on which every night
a wood fire was kindled.   A pious hermit was entrusted
with the care of the fire, and had in return the right to demand two-pence sterling of every passing vessel.   To enable him to serve his heavenly Master while performing
his earthly duty, a little chapel was built near-by in honor
of the Virgin Mary.   Soon the hermit was able to hire
assistants ; a few fishermen joined the company, and ere
long a little village sprang up around the diminutive
light-house.   In the course of time the little island was
literally swallowed up by the voracious ocean ; the rocks
on which the houses had stood disappeared one by one,
and with them the cottages ; the chapel also, and the very
ruins of the original light-house, have passed away; and
now no trace of it all is left except a bare rock and a few
sand-bars, barely visible at low tide.

A new tower was ordered to be built in 1584, and the
king employed for that purpose a famous architect from
Paris, Louis de Foix, who subsequently built the great
Escurial for Philip II.   The father died before the work
was completed, but his son finished it in 1610.   The new
tower was, with the lantern, 70 feet high, and divided
into four stories, the second of which was proudly called

the king's apartment, and adorned with butts, costly hangings, and elegant sculptures. Above it was a chapel of fair dimensions. So far the original work of the two Foix is still in existence, having braved the storms and the waves for more than two centuries. The upper part, however, was gradually changed, and at last entirely destroyed, to make room for the present superb structure. Already, under Louis XV., in 1727, an iron lantern was substituted for the lantern in brickwork, which had long answered well, but was now objected to, partly because the stones had become calcined by the constant effect of the fire, and partly because the solid masonry obstructed largely the light. Then it was found that the insufficient height did not allow the light to be seen at a greater distance, and the project was formed to raise the upper part some sixty feet higher. This plan, which was not without danger, was successfully carried out in 1789, by Teulère, chief engineer at Bordeaux, and made his name as famous as that of the first builder.

The present structure is perfectly plain on the outside, presenting to the wind and the waves everywhere a smooth surface on which they can find no corner, no opening, no sharp edge even, on which to take hold. But there is majesty in this very simplicity of the great monument, as it rises boldly from the bosom of the ocean. It stands out grandly against the clear western horizon, and yet it appears under a hundred different aspects. On a bright sunny day it shines brilliantly in dazzling splendor; at other times it seems to float vaguely and dimly amidst shapeless vapors. At night it breaks out all of a sudden with its beautiful red light, and with its gigantic eye seems to scan the wide waste of waters, to warn the imprudent and to cheer the despairing.

This patriarch of light-houses has, besides, had the honor of being used for trials of all the great inventions that are made from time to time in the manner of lighting the beacons. Thus it was here that Tresnel established one of the very first of his dioptric apparatus; as here also, ages ago, coal had first been substituted for wood, and then lamps for a fire. In 1782 not less than 80 lamps, each accompanied by its reflector, nearly filled the lantern. More recently still greater improvements have been added, and even now the question is agitated of introducing here also the electric light, which has already done such signal service in the harbor of Havre. Nor must we overlook the remarkable fact that the tower of Cordonan possesses the peculiar power of resistance with which high light-houses are endowed, after the manner of giant trees. It does not mind the masses of rain, of hail, and of snow, which the fierce winds drive furiously against its sides and the solid panes of its beautiful lantern. It stands calm and undisturbed amid the host of gigantic waves, which in times of tempest come up in unbroken ranks and dash their heads against the granite walls; they glide smoothly from the polished sides, break into foam against the galleries, and, after deluging the cupola on high, run madly off into the Bay, but not without having exacted a tribute from the proud granite. The tower bends imperceptibly to do homage to the terrible power of its adversaries. The keepers assert that during violent storms the vessels filled with oil, which are kept in one of the topmost chambers, show a change of level of more than an inch—which implies, that the tower must describe an arc of more than three feet! The same experience has been made in other light-houses; as soon as the structure rises to a height of 120 feet, these rockings become sufficiently strong to spill liquids kept in shallow vessels, and to make the weights of the clock-work strike against the inner sides of the pipes to which they are confined. On a much larger scale yet, the steeple of the Cathedral of Strasburg, 440 feet high, bends under the fury of storms, and rocks high in the air its long ogives, its delicate columns, and the gigantic cross that crowns the summit.

If the tower of Cordonan is the very oldest of the French light-houses, and well known to mariners of every nation by its bold position and the great benefits it has bestowed upon countless vessels, little has been heard as yet of the youngest of the family in far distant waters. At the other end of the world, in the South Pacific Ocean, there lies an island famous for its coral reefs, which connect such a number of islets, rocks, and sand-banks, that the navigation is rendered most intricate and dangerous, and the land can be approached by two open-

ings only. The French government, to protect its rapidly increasing commerce in those regions, ordered a
light-house to be erected on a sand-shoal far out in the
open sea, but sheltered against the violence of the sea by
coral reefs on three sides. This comparatively safe position encouraged the builders to select cast-iron for the
structure.* The tower, built in the city of Paris, consists
of a skeleton, covered with plates of iron and fastened
together with strong iron rivets, the whole building
being anchored as is were to the sand itself by cast-iron
screws. In spite of the light and graceful appearance
which is thus given to the building, and which attracted
the admiration of all beholders at the last Exposition at
Paris in a twin structure, the tower is so strong that the
rockings perceived in solid structures of stone and brickwork are here hardly noticed. The height up to the
lantern amounts to 165 feet, and the admirable apparatus
for lighting sheds a fixed white light to a distance of
twenty-two miles. When the tower, sent out piecemeal
in 1865 to its far distant home, reached the island, and
was inaugurated, amid solemn ceremonies befitting the
occasion, the motley crowd of sailors, soldiers, and
French colonists nearly disappeared amid overwhelming
numbers of savage natives, whom the splendor of the
display had attracted from all directions. But as soon
as the beneficent light had begun to shine upon the dark
waters, the light of civilization also dawned upon the
hearts of the poor outcasts, and soon cannibalism ceased,

peace brought prosperity, and the blessings of our faith were brought to these remote regions by zealous missionaries.

While light-houses shine on the coast of every civilized country, marking every cliff and reef on the rockbound shores of Europe, and warning sailors against the countless shallows and dangerous currents on our own coast, smaller lights are placed at the entrance of harbors, and on places which by their nature render the erection of larger structures impossible. This is done by means of light-ships riding at anchor, before treacherous sand-banks, rapid currents, submarine whirlpools, or reefs which the tide covers completely at certain hours of the day. The invention is said to be due to a remarkable barber of Lynn, in England, called Robert Hamblin. His good looks brought him, with the hand of a ship-owner's daughter, also the right of property in a small vessel, in which he was in the habit of carrying coal from place to place along the coast. Thus he became personally familiar with the great danger of the Goodwin Sands, which, by their shifting, changing nature, have perhaps proved more fatal to life and property than any other quicksands known in the great ocean. Chance brought him in contact with a man of great ingenuity but small means, David Avery, who suggested to his new friend the expediency of putting a light not on a tall tower, which was utterly beyond their means, but on board an old hulk unfit for any other purpose. The energy of one partner and the skill of the other combined to erect a floating light at the Nore, near the mouth of the Thames, and, to support the enterprise, they took it upon themselves to levy a small tax on all passing-vessels.

The novelty created great displeasure at Trinity House; they disliked seeing outsiders originate a new plan, the success of which could not be denied, and they objected still more to the levying of duties by any one but themselves. When therefore Avery, encouraged by the approbation he had earned by all sailors, openly announced his intention to place a similar light-ship in the waters near the Scilly Islands, complaint was made before the Lords of the Admiralty against the intruder. Unable to obtain a hearing, the Corporation appealed directly to the King, setting forth the lawlessness of men who ventured to tax vessels by their own authority, and succeeded here so well, that Nore lightship was ordered to be removed. Avery, threatened with the loss of all he had and all he had hoped still to earn by his happy invention, thereupon entered into negotiations with the indignant members of Trinity House, and pleading the value of his idea and the heavy outlay which he had incurred, induced them to agree to a compromise. He surrendered all his rights to the Corporation, and received in return a lease for sixty-one years, in payment of an annual rent of a hundred pounds.

This was the origin of light-ships; a new evidence of

the great ability which official authorities display in every age to discourage useful inventions; and, on the other hand, of the unfailing success of men who combine energy with perseverance. Trinity House made light of Avery's suggestion; it declared that Sir John Clayton had had the same idea fifty years before—but for all that Avery and Hamblin were the first to try it practically and to succeed.

A light-ship is a painful sight to a sailor, as far as its appearance is concerned. The hulk is, of course, like all others; but the short, thick masts have no yards nor sails, and bear on the head large balls of wood. The very essence of a vessel, its easy, graceful movements, its readiness to yield to a slight pressure of the hand on the rudder, or a gentle puff of wind on the sails, is missing. The only question here is not motion, but immobility; how to remain stationary in spite of the elements—this is the first requisite in a light-ship. Amid the greatest violence of the waves, during the highest spring tides, and in the very centre of powerful currents, the stout vessel must ride quietly at anchor, undisturbed by all the turmoil and uproar around it. Like a galley-slave of old, it is fastened by iron chains and huge cables to the spot, where henceforth it is to remain immovable. The length of these chains is sometimes enormous; the light-ship at Seven Stones, riding in waters 270 feet deep, is fastened to a chain a quarter of a mile long, and besides secured by a number of stays and shackles. The cases

of light-ships being torn from their moorings are very
rare; when the elements become too strong for the
fastenings, and the light drifting about at random might
mislead ships in the neighborhood, a red signal is immediately hoisted, a gun is fired at prescribed intervals, and
generally the vessel is brought back to its place in an
incredibly short time. But the calm courage and the
presence of mind required on such occasions speak highly
for the character of the crews of these vessels in general,
since none has ever been lost, nor has it been ascertained
that any calamity has ever occurred in consequence of
a light-ship being torn from its fastenings. To make
assurance doubly sure, however, these is generally a
vessel held in reserve at a short distance from the light-
ship, which can be telegraphed for, and may at any
moment be sent to occupy the vacant station for a time.
The light-ships of Trinity House are painted red, those
of Scotland black, while our own are generally adorned
with longitudinal stripes of different colors. On the
sides of the vessel the name is commonly painted in
gigantic letters, and a special flag is assigned them in all
maritime countries.

The floating lights of England and those of the Union
amount to the same number—forty-seven. American
light-ships have only of late come up to the standard of
foreign vessels of the kind, and a considerable number
of them are so placed that in times of stormy weather
they can leave their post and seek refuge in calmer

waters. France has the smallest number, partly because the coasts require less, and partly because certain positions, like those in the Bay of Biscay, are such as to have so far defied all efforts to establish a light-ship in them permanently.

The crew of larger light-ships consists in England of a master, a mate, and nine men. Three of these attend to the lights, while the other six, among whom there must always be an able carpenter, keep the vessel clean and in order. One-third of the crew is absent, being allowed to live on shore during their time, since experience has taught the authorities that an unbroken stay on board is fatal to health and morality alike. The terrible monotony of such a life, which gives no occupation to the senses or to the mind, and the incessant, silent struggle with the power of the elements, overwhelm the mind, induce apathy and melancholy, and lead to crimes or to suicides. Not without reason, therefore, did Dickens include the crews of light-ships among his curiosities of mankind. Hence the master and the mate alternate by months, while the crew spends two months on board and one on shore. But the elements do not always permit the regular change, and many a crew has been compelled to remain for months and months without any intercourse with the land. In olden times, it is said, more than once the men were thus doomed to death by starvation. It is touching to hear how the poor fellows, when on shore, dream incessantly of the sea, while

on board ship their minds wander as faithfully back to
the shore.

The lanterns, in which the lamps are enclosed, are
placed around the top of the mast; during daytime they
are lowered in order to be cleaned and filled with oil;
at night this crown of lights is raised again and shines
to a great distance. Every vessel is besides provided
with a gun and a gong, and thus they are enabled to be
of most valuable service. The famous light-ship near the
Scilly Islands has witnessed but two shipwrecks on the
reefs near which it has anchored; in one case the crew
saved one man, in the other all the passengers except the
wife of a missionary. The Government does not encour-
age these humane efforts by rewards, as the sole duty of
the crew is to maintain the light, which the men cannot
be allowed to neglect for a moment, since while
one life may be saved, a hundred lives may be imper-
illed. The light-ship at Seven Stones occupies the most
exposed position of all such vessels; nevertheless it has
never been torn from its moorings, although the waves
frequently wash over the deck for hours, and when the
sea strikes the sides, it "makes a noise like a four-
pounder," as the captain says. English crews are always
taken from the navy; in our country the keeper is generally a civilian, and frequently hires his men. Hence
the discipline is not as strictly kept up with us as in
England, where a man was once on the point of losing
his life because he had secretly left his light-ship in order

to attend the funeral of his wife, who had died in London. He was pardoned, however, and escaped with a severe reprimand.

Where neither light-houses can be built nor light-ships moored in the sea, to serve as guiding stars to mariners, there sailors look instinctively for landmarks to direct them in their navigation. Now it is a steeple, and now a windmill; in lonely countries a single tree or a prominent rock, which serve thus as guides. Isolated mountains, like the Peak of Teneriffe, volcanoes throwing off their clouds of smoke, or distant glaciers shining afar off in the bright light, are gigantic landmarks of this kind, welcomed by the sailor from distant seas. Among these remarkable points some have acquired historic fame. The most ancient of all are the Pillars of Hercules, of which a French poet once sang that there was but one thing wanting to their glory—that they should have existed. Hesychius, to be sure, mentions three or four pillars which stood near the Straits of Gibraltar, and were reported to have been placed there by Hercules; and the great Arab Geographer, Edrisi, speaks of six, of which one was placed as far east as Cadiz in Andalusia, while the westernmost stood amid the islands of the Dark Sea, the Canaries, and bore the inscription : *Ne plus ultra!* But Strabo already expresses his doubts of their existence when speaking of Cadiz; and it is certain that besides the allusions we have mentioned, there exists no evidence of their existence.

Another famous monument, long known as a landmark, is the beautiful pillar that stands in Alexandria, and is known under the name of Pompey's Pillar. It is the first object that becomes visible to those who approach Egypt from the sea, and towers apparently high above the obelisks and minarets of the eastern city. Its precise height was unknown until the French expedition came to Egypt in 1798, when some of the savans attached to Bonaparte's army undertook for the first time to obtain its measure. They resorted, for this purpose, to a very ingenious device. A huge kite was started, to which a long rope had been fastened; when the kite was floating tranquilly in the air just above the top of the column, it was suddenly pulled down by the rope, which thus was stretched over the capital as upon the outside of a pully. Then a larger rope was substituted for the first, and fastened with stakes to the ground at the foot of the pillar; on this a child went up to the top and fastened there, by means of a block and fall, a sufficiently strong tackle by which the savans were hoisted up in a chair.

The beautiful column consists of four huge blocks of rose-colored granite, but the shaft alone is of undoubted antiquity and in the purest style; the capital and the pedestal have evidently been added at a later period and are not in proportion. Nevertheless the column gains by its isolation, which makes it appear higher than it really is, and by the exquisite beauty of the Corinthian capital, more than it would have gained if the whole had been completed like the shaft, in the Doric order. According to the testimony of many authors, the present pillar did not always stand alone. Arabic authors speak of it as being surrounded by an immense portico, whence it was called in Arabic *Amouad el Saouary*, the Pillar of Pillars, a name which was by ignorance corrupted into the Pillar of Severus. Captain G. H. Smyth thinks, on the contrary, that this is the column of which Aphthonius says, that " it bears the elements of all things," an expression which might have referred to the great copper disc of which Hipparchus makes mention. As another Arabic author positively asserts that he saw, in 1200, a cupola on top, many have inferred that the column may have been intended for astronomical observations, and perhaps stood in the very centre of that magnificent Serapeum, which is thought by some to have held the magnificent library of Alexandria.

Others again have looked upon the column as from the beginning intended only for a monument, and have ascribed it to Cleopatra, who erected it in honor of Pompey. Although it still bears the name of the triumvir, none of the old authors who have described Egypt in their pages, Pliny, Diodorus of Sicily, or Strabo, mentions the pillar, which they would certainly have done if it had existed in their days. Pocock thinks it was erected in honor of Titus, or Hadrian, while Abul Feda attributes it to the Emperor Severus. The difficulty seems only to have been increased by an inscription which was discovered by Pocock. While examining the column carefully and measuring its dimensions, he had noticed in the strong sunlight, between eleven o'clock and noon, traces of a Greek inscription on the western side of the plinth of the base. Numerous blank spaces and the vague outline of the letters prevented him from deciphering the words, but at last several French and English savans succeeded in unravelling the mystery. It was found to read thus:

TO

DIOCLETIAN AUGUSTUS,

MOST ADORABLE EMPEROR,

TUTELARY DIVINITY OF EGYPT,

I, Po. . . . .   PREFECT OF EGYPT,

CONSECRATE THIS.

It is noticeable that the foundations of the column are formed in the coarsest manner, stones of every kind and of all dimensions being heaped up together. Efforts have been made from time to time to dig under these foundations in search of hidden treasures, and this has been done so recklessly, that the pillar now inclines more than seven inches to the west. Nor are the usual efforts wanting to disfigure the beautiful monument for the benefit of ruthless travellers, and towards its upper part, the graceful shaft is defaced by the names of some English tourists, written with pitch in letters of ten feet high.

Gradually the eminent services of such landmarks were more and more appreciated, and it became customary to erect them on the coasts, where none had been provided by Nature. The Etruscans are said to have been the first to erect little pillars for the guidance of mariners by day, as light-houses served by night, and in our day the good old custom has been systematically revived. Small pillars of stone or structures of timber are erected on prominent positions and painted white, where they appear against a dark background; brown, when they are seen against the sky. If it appears desirable to mark a submarine reef, on which vessels without a good pilot might be stranded, beacons and buoys are used, which have on the top a cash, or a ball, or even a larger structure composed of iron bars. As these buoys, however, are useful only in clear weather, when they can be distinctly seen, and cease to be of service in fogs, they have frequently been provided with large bells fastened inside of a small open-work cage, and set in motion by the agitated waves. New additions have quite recently been made to these warning sentinels, which abound in all navigable waters from the wide waste of the ocean up to the head waters of great inland streams. In France, for instance, all buoys which are to be left on the larboard in coming from outside, are painted red with a white crown on the summit, while those that are to be left on starboard, are painted black. Others are adorned with stripes, lozenges, or squares, in all possible colors; and some actually possess mirrors, in which they reflect the rays of the sun, or the light that comes from light-houses at a little distance. On our own coast, where fogs are very frequent and currents near shore very dangerous, no expense has been spared, and almost all difficult places which are frequented by vessels, are provided with large bells, which are rung during fogs at fixed intervals, or with enormous whistles, blown by an ingenious apparatus. Even here the Etruscans are said to have taken the lead ages before our own invention, although in those days the means were of the simplest. A rude conch-shell, such as is still in use on West India plantations, was blown from the coast by men stationed there during dark weather; and later, marine trumpets were invented to take their place.

These were primitive ways certainly in comparison with the ingenious contrivances by which in our day vessels afar off are informed of all that is important for them to know. At every port and harbor there are stations, from which by means of flags or balls covered with rope-nets, telegraphic news can be sent out to a distance of many miles. The ship that comes home from its long voyage, sees thus at a glance how much water there is in the channel through which lies the entrance to the desired haven. Every item of importance is thus signalled out by a number of flags and their combinations, while at night lights take their place. If we add to this, that in France, as in fact in most maritime countries, the laws prescribe that every vessel entering into port must carry lights, we can readily understand how well the whole coast of civilized countries is lighted up, and how brilliant the scene is at the entrance of great harbors. This custom also was known to the ancient Greeks already, and graphic descriptions are found in old authors of the bright lights that shone from their beautiful vessels, the height on the mast indicating the rank of the commander. The common usage of our day is for every vessel, even out at sea, to have a white light on the mizzenmast, a red one on the larboard, and a green one on the starboard side. Smaller vessels use a wick dipped in turpentine in a brass vessel, which sheds a bright light, and burns all the better when it is wetted by rain. Thus alone could the number of dangerous meetings in dark nights be diminished, and in the course of a few years thousands of human lives are saved by this simple precaution.

It is sad to have to add, that light-houses and beacons have not escaped the fate of all benevolent measures, but have served from oldest times down to our own, in the hands of wicked men, to ruin others and to drive them into destruction. While in one place the welcome light shone brightly, and seemed to say to despairing hearts: Come safely on, here hearts beat for you, eyes are watching and hands waiting to receive you! a few miles further on, other lights were kindled to draw the unwary upon reefs and shallows, and to enrich the spoilers with the wealth of the unfortunate shipwrecked. The annals of antiquity are full of stories of such infamous wreckers, but thinly disguised under the half fabulous guise of fair sirens or one-eyed cyclopes, of a beautiful Circe or cruel Cushite priestesses on the shores of Campania. Do we not all remember the stratagem of king Nauplius, who ruled over the island of Euboea? Irritated against the Greeks, who had murdered his son Palamedes, he kindled large fires on Mount Caphareus—now Kaoo Daro—to attract the Greek fleet during a storm to its dangerous rocks, and not a vessel escaped!

During the whole of the Middle Ages the poor ship-wrecked mariner was looked upon as a God-send by the people on the coast, and his ship was plundered according to law and custom. " Good are the wrecks that come to the king's coast; they are the king's," said the old French law. Severe decrees, it is true, were issued against those inhuman wretches who lighted up decoy-fires for the purpose of leading mariners into dangerous places and then profiting by their misfortune. " They shall be put into the sea and plunged till they are half-dead, then pulled out and stoned and knocked down like dogs or wolves," say the same laws. But what was the result? As late as 1794 a formidable reef on the coast of Brittany was looked upon as a gold-mine by the inhabitants. " While the honest man trembles at the sight of the danger," says a French author of that day, " the pitiless dweller on that coast arms himself with sticks and ropes, hides behind the rocks, and pounces upon all he can seize without being caught by the police. Formerly he killed the unfortunate man who stretched out his arms towards him from the waters, imploring his help; now he has become more humane; he leaves him his life, but nothing else. If the police interferes, they become furious; their wives come to their assistance, and united they fall upon the gendarmes; shots are fired, blood flows, and lives are lost. It is the height of injustice and of cruelty in their eyes for these men to deprive them of what God has sent them in His goodness."

It is not less curious to see how the same cunning devices have been used from time immemorial for the purpose of misleading ships. On the coast of the Black Sea, already in the days when the ancients yet called it the Inhospitable Sea, barbarians had learnt to tie a lantern between the horns of a cow, and then to lead her along the edge of the water, knowing that the motion thus given to the light closely resembled a light on board ship; and the same infamous custom prevailed in the last century yet on the coast of France. Nor is England exempt from such outrages. In 1825 a Greek brig, the "Cimoni," was wrecked near Alderney, and so completely plundered that the crew was literally left without a single article of clothing by the civilization of the nineteenth century! As late as 1866 the "Morning Post" complained that, contrary to the general impression, the barbarous custom of kindling decoy-

fires on the coast was yet in existence in some parts of
Cornwall, and even in Durham.   The unusually large
number of vessels which had at that period been lost
between Sunderland and Tynemouth, on rocks on which
lights had been seen burning, gave color to this assertion.   But worse is still behind.   While England can
boast of her admirable life-boats, which have been the
means of preserving no less than eleven thousand four
hundred men from a watery grave, regular associations
of wreckers have been discovered to exist in some
counties.   These men, instead of kindling decoy-lights
and plundering the crews of vessels which they have led
into destruction, go out to seek their spoil on the open
sea.   As soon as a small vessel is signalled from the look-
outs they keep, they jump into their cutters and pull out
to the ship; they board her, and without ceremony take
possession of her to carry her into the next English harbor.   Here they swear that they have rescued her from
sinking, and demand the usual salvage.   Instances of
similar atrocity, it is well known, have occurred on our
own coast in the neighborhood of the Florida Keys, and
it is to be apprehended that no nation is entirely free
from such disgraceful abuses.

Nor must we overlook the fact, that light-houses themselves, and the most perfect lights they can boast of, are
often the cause of great disasters.   The latter happened
to be the case not long ago, when an emigrant ship, the
" Dunbar," went ashore on the Sidney Headlands, just

underneath the superb dioptric light which burned on the summit, but did not suffice to show the dangers immediately at its foot. On the other hand, it must be borne in mind, that the lighted headlands and sands are the true points of danger on every coast, and that if light-houses sometimes are the direct cause of shipwreck, this arises from the fact that the seaman must first see the danger before he can avoid it. This has led a member of Trinity House to propose what he calls his Fair Way—a series of light-ships, of the simplest construction, all the way up the Channel—so that a ship making the westernmost of the Lizard would be enabled to make her way up the mid channel in perfect safety. For, after all, Man is a match for Nature. Whilst the ocean fights against the mariner and hurls him on the coast with relentless fury, Art, from the land, replies by her cunning engines, and wrestles with the waves for the stake of human life. Thanks be to God, the victory is not to the strongest, nor the race to the swift; and year by year the number of lives lost on our coast is diminished by the noble efforts made to prevent shipwreck, and Light has been found, at sea as on land, the most efficient agent in securing life and property alike.

IX

LIGHT-HOUSE STORIES.

"Hail, Holy Light! Offspring of Heaven's first-born."—MILTON.

THE traveller who sails up that most delightful of all streams, the Rhine, sees on his right hand, when not far from Bingen, a pretty toy-castle, which raises its perfect battlements high above rocks and mountains. On one of its tiny turrets floats the black-and-white banner, with the fierce Prussian eagle in the centre; for the seat and the occupation of the robber-knight of old have both been assumed, though on a royal scale, by a prince of the house of Hohenzollern. If it is the good fortune of the traveller to ascend that part of the river in the sweet twilight of an autumn evening, he will soon after sunset see a strange reddish flame blaze up near one of the smaller towers; it hangs apparently free in the air, but nearly over the bank of the river, and sheds its ruddy glare up and down the dark waters. As he turns round the tiny promontory, which serves as a gate to the long, open stretch on which the fire shines, he discovers at last that there is a quaint iron basket, fastened by huge iron rods to the stones of the tower, and that inside the grating large logs are smouldering slowly in the damp night-air. As he looks down on the dark waters, with their strange red glow on every wave and the wash on the bank, he perceives here and there enormous blocks of stone nearly rising to the surface, which threaten the little skiffs with destruction, and are formidable even to steamers ; and now he understands the friendly meaning of the warning fire on high.

He has seen here, in the heart of Europe, in the middle of the nineteenth century, the precise form and shape of the most ancient light-house that is known to our annals.

So, at least, we judge from the records left us in many a parchment and the designs cut on ancient medals. For the light-houses of antiquity have, unfortunately, crumbled into dust and débris with the Roman Empire itself, and all that we know of them we have gathered painfully from the numerous but vague descriptions of their contemporaries. As with many other things, so here also we would willingly exchange the many words for a few stones. The science and the ingenuity of a Rawlinson and a Layard would have read more in a handful of carved rocks that once belonged to the foundation of an ancient light-house than we can learn from countless pages written on the subject.

The Greeks attributed the first structures of the kind, almost as a matter of course, to their favorite Hercules,

whose greatest labor must have been to bear the burden
of all the wondrous things he was said to have accomplished. But even he can hardly have been thought the
builder of the numerous beautiful towers raised by the
Libyans and the Cushites, who dwelt in the provinces of
Lower Egypt, for the purpose of bearing great fires on
their summits. Guiding stars in the night, they served
in the day-time as points of observation; and many are
the weighty facts of astronomy, on which all our knowledge of the universe depends, that were here first ascertained and recorded by the sages of antiquity. But as
knowledge was in those days not only power, as with us,
but worshipped as divine, these famous towers were temples also, and bore each the name of some great divinity,
while grateful sailors, rescued from danger and death,
enriched them with their votive offerings. Modern speculation has added still another attraction to these mysterious buildings—it looks upon them as depositories of
all the geographical knowledge possessed by the ancients,
where maps of the coast and charts of the navigation of
the Nile were preserved, first simply drawn upon the
walls of the building, and afterward transferred to
papyrus rolls. It was thus that those temples were
transformed into learned schools, and the priests changed
into teachers, who imparted the knowledge of hydrography, and taught the art of sailing vessels by the
guidance of the stars. These venerable towers were
therefore light-houses in more than one sense. Within,

the bright light of knowledge was diffused by zealous priests and learned sages, to go forth to all the nations that then navigated the one great sea of the civilized world—the Mediterranean. Without, a machine of iron or bronze, consisting of three or four branches in the shape of a dolphin, or some other marine animal, and connected with each other by garlands of beautiful foliage, contained large masses of fuel, which were faithfully watched over and renewed during the dark nights. A long iron bar of great strength, moving on a hinge so as to enable the priest to draw the colossal brazier to him, supported the bronze basket. The seas then swarmed with small vessels; and as each one of these also bore its fiery signal on the bow, to avoid disasters by night, and to show by its size and its height on the vessel the rank of the owner, fire greeted fire with delight, and the whole scene must have been one of great beauty and interest.

What these early light-houses were called is a matter of great dispute among the savans of our day, but does not, after all, matter much for practical purposes. Some, it is said, were named Tor by the Libyans; others, which occupied the highest eminence within the walls of a city, bore the name of Bosrah, a title which was afterward transferred to the citadel of Carthage. When they were situated in the open country they were generally built in the form of round towers, and then known as Tith; and the pretty legend was long current that the myth of the Cyclopes, killed by the shafts of the sun-god Apollo, meant nothing more than the manner in which the lights that burned on the Cyclopean towers along the south coast of Sicily were extinguished by the rays of the rising sun.

The first regular light-house which is even honored with the supposition that it had already a revolving light, is one represented in the Iliac Table and ascribed to the Ninth Olympiad. Its fame, however, dimmed by the remoteness of its existence, was entirely eclipsed by a later one, which has given its name in French and other languages to the whole class of similar buildings, even as Columbus lost the glory of leaving his name to our Continent. This tower stood on the island of Pharos, near Alexandria, in Egypt, and became subsequently the model after which all structures of the kind were built for many centuries. Such was the case, we know with certainty, when poor old Claudius built the famous tower at Ostia, which seems to have been the most beautiful among the many that lighted up the coasts of Italy. And yet Rome seems to have hung out her shining beacons with the same solid splendor that characterized all her noble structures at home and in the provinces; for we read in Pliny of the superb towers of Puteoli and Ravenna, and we know all about the great light-house at Messina, which gave its name to the straits between Italy and Sicily, where the far-famed rocks of Scylla and Charybdis were still the terror of sea-faring men. The magnificent temple, finally, which Tiberius in one of his caprices built in the very midst of his twelve magnificent villas on Capri, was one of almost fairy-like beauty, and with its grand blazing fire lighted up the sea for miles and miles, so that the poor fishermen of the islands began to dread its weird splendor, as if it dared to defy the gods themselves, and believed more than ever in its magic nature when an earthquake levelled it to the ground, a few days only before the death of the terrible tyrant.

How sadly even then already the benevolent efforts of wise and sagacious men were defeated by the wickedness of others we learn from the description of another celebrated light-house, which stood on a lofty promontory where the river Chrysorrhoas threw itself into the Thracian Bosphorus. "At the top of the hill," says Dionysius the Byzantine, "around the base of which the river flows, stands the tower Timæus, of marvellous height, from whence one overlooks a vast expanse of water, and which has been built for the purpose of insuring the safety of those who sail on it, by kindling large fires on the summit for their guidance. This was all the more necessary as there were no harbors on either side, and anchors could find no bottom on which to fasten their flukes. But the barbarians along the coast lit other fires at the highest parts of the shore, in order to deceive the sailors and to profit by their shipwreck. Now the tower is in ruins, and no light shines any more from its summit."

We know but little of the precise form of these ancient light-houses. Herodian, it is true, asserts that they were built in the same manner as the catafalques of the emperors, but the latter were square constructions, adorned on four sides with paintings and sculptures, while the light-houses were, at least in many cases, built in the shape of round towers. As such they appear on the only two ancient coins or medals on which a Roman port with a Pharos is represented. In both instances the latter consists of a round structure of massive stone, rising in four stories, diminishing toward the top, and crowned on the summit with a blazing fire.

More is known of the great Pharos itself—for so it soon was called universally—which Ptolemæus Philadelphus is said to have built on the tiny island of that name which lies in the shallow waters near Alexandria; for it became so famous in times of antiquity by its colossal size and magnificence of ornament that it was placed among the Seven Wonders of the World. A few old writers, it is true, are gallant enough to ascribe the beautiful building to the good taste and wise foresight of the dusky queen, Cleopatra, the Mary Stuart of antiquity; but modern authorities are little influenced by deference to the sex, and stoutly deny her claims to such a distinction. A good Benedictine monk, Dom Bernard of Montfaucon, adds still another romance to the famous tower, and recounts how the ingenious builder of the tower, Sostrates, succeeded by a clever stratagem in handing down his own name to posterity, while that of the great king, for whom he acted as architect, became dim and doubtful in succeeding ages. He cut, it is said, the words "Sostrates of Cnidus, son of Dexiphanes; to the gods who save sea-faring men," deep into the hard stone on the face of the temple, and then covered the inscription with a slight coating of perishable material, on which the name of King Ptolemæus was written in gigantic characters. The coating and the name fell off in a short time under the influence of wind and weather, and nothing was seen but the legend that gave all the glory to Sostrates.

The story, if not true, is well devised, as the Italians say, and has found ready believers in all ages, few men being willing to admit that even among sovereigns such modesty could be found as would induce them voluntarily to relinquish the gratitude of posterity in favor of a mere servant. Other savans of our day, and among them men of the highest authority, like Champollion, have tried to escape from the dilemma by giving the honor to another Ptolemy; but we are disposed to agree with Edrisi's quaint but solemn conviction: "God alone knows the truth of the fact."

The tower itself stood upon a little island, the site of which is now covered with the buildings of the modern city of Alexandria. In those days, however, the island and the town were nearly a quarter of a mile apart, a distance which Homer poetically enlarges to a day's journey from Egypt. At a later period the island was connected with the main land by a long causeway and a magnificent bridge. According to the minute but very obscure descriptions which we find here and there scattered in the works of ancient authors, the tower consisted, like Babel itself, of several vaulted stories, and, if we are to believe Pliny, the cost of erecting it amounted to the almost fabulous sum of eight hundred talents.

The very fact of its existence, however, has been doubted; how much more uncertain, then, must be its identity? It may have continued as late as the twelfth century, since Edrisi, the famous geographer of Nubia, who united in his person the rare lore of the Arabs with the gentle science of the Sicilian capital, has seen it. "This Pharos," he says "has not its like in the world, as far as its construction and its solidity are concerned; for, independently of the fact that it is built of an excellent kind of stone, the courses of these blocks are joined to each other by means of molten lead, and the joints are so closely adherent to each other that the whole is impervious, although the waves of the sea beat on the northern side incessantly against the building. The ascent to the top is made by a staircase built in the interior, and as wide as ordinary stairs are in other towers. But the steps only go half way up the monument, and there the building becomes, from the four sides, narrower than below. In the interior, and under the staircase, there are several rooms. From the gallery upward the light house rises straight to the summit, becoming smaller and smaller, until at the top a man can span it with his arms. From this same gallery you ascend by means of a second staircase, but of much smaller dimensions than the lower one, and lighted by means of small windows in the outer wall, so as to give light to the persons who ascend, and to enable them to place their feet securely on the step."

The fire was kept burning continually, appearing by night like a brilliant star, visible to the enormous distance of nearly a hundred miles, and rising by day in the shape of a dark cloud to the heavens. This resemblance to a star seems to have been as fatal in those days as it still is in our time; for Edrisi says that many sailors had mistaken the fire for a well-known star, and directed their course accordingly, in consequence of which they had wrecked their vessels on the sand-spits near the shore. To avoid similar errors in our day many light-houses are provided with two lights, one above the other, so that neither can be mistaken for a constellation.

The credulous Benedictine, whom we have quoted before, has his goodly store of romances in connection with this great light-house also. He had learned from Arabs and sailors of other nations that, according to popular tradition, Sostrates built the colossal tower, for greater safety, on four immense crabs of glass! Nor is the monk alone in his statement, for greater authorities also repeat the same story on the faith of an ancient manuscript, which pretends to give an authentic account of the Seven Wonders, and was actually studied by the learned Voss. Another marvellous story connected with the building, and long faithfully believed in, is the report that Alexander the Great caused a mirror to be placed on top of the tower, which was constructed with such wonderful art that it showed on its highly polished surface every object at a distance of more than a hundred miles, and thus enabled him to recognize the hostile fleets that came to attack Egypt days before their actual arrival. The mirror, it was added, was destroyed by a daring Greek, who availed himself for the purpose of an opportune moment when the whole garrison was asleep. The only difficulty in the way is the fact that the great Pharos had not yet been built in the days of Alexander, and hence the good Benedictine winds up his account of this tale with the words: "It is rather in the genius of Orientals to invent such unreasonably marvellous things."

There is but one other light-house mentioned in the annals of antiquity of equal interest with the Alexandrian tower. Roman writers of indisputable authority tell us that when the mad Emperor Caligula returned from his fantastic expedition into England, which never went further than to the shores of the Channel, and resulted in the picking up of a few shells, and laying these spoils of the ocean at their commander's feet, he ordered a light-house to be erected in honor of the fictitious victory, to guide vessels by night into the harbor that had been the scene of the glorious exploit. The place became subsequently more and more famous, and in the days of the great Napoleon was once more the scene of a vast assemblage of troops ready to invade England and to conquer the kingdom.

This famous tower of Boulogne shed its light for centuries over the stormy waves of the Channel. Already in 191 it was revered for its blessed influence; and Commodus caused a medal to be struck on which the light-house and the departure of a Roman fleet appear, in company with his victorious title of Britannicus. Its prestige continued as long as Boulogne remained the favorite place of embarkation for all the Roman troops that went over from Gaul to Britain. It appears next in the annals of Charlemagne, whose wise policy neglected no means of enlightenment, from the material fire on light-house and beacon to the spiritual light which was diffused by the countless schools he endowed in his vast empire. The place grew, and became in course of time a fortified place, as important by its vast works of fortification as by its natural position, which commands the Channel in front and the two banks of the little river Liane, which there falls into the Channel. The good people of the neighborhood were so deeply impressed with the grandeur of the wonderful tower, and especially with its great height, that they stood in constant fear lest the lightning from heaven should destroy it, as it had done with the tower of Babel. But its final destruction came from the carelessness of the very men who

were proudest of its magnificence. Although the sea
beat incessantly against the foundation, and at the time
of high tides even against the sides of the tower, no protection was ever raised toward the Channel; a number
of springs, besides, worked underground, and undermined
the structure slowly but surely ; and, as if these agencies
had not been threatening enough, large quarries were
opened in the very hill on which the light-house rested,
till at last the fortress, the tower, and the very cliff on
which the whole had been erected, fell one fine day and
tumbled into the sea. The catastrophe was followed by
a most ludicrous lawsuit between the lord of the soil and
the town of Boulogne, which had heretofore paid him a
certain rental. As the soil had disappeared, the citizens
considered themselves relieved of all obligations toward
the owner; but the latter carried the suit up to the Royal
Parliament, which in 1656 condemned the city either to
pay, as heretofore, two thousand herrings annually, or
to restore the place to its former condition. As such a
restoration was not exactly in their power, it seems that
they paid the herrings down to the days of the French
Revolution. The tower has in recent times been replaced by an elegant light-house with several lights ;
and though less famous than in days of old, it still renders eminent service to the numerous vessels that nightly
pass the populous town.

Nearly opposite to the work of Caligula there rose,
near Dover, a sister tower, built like the former by the

hands of Romans, and like it destined to perish ingloriously by neglect and false economy. Its very place is uncertain, as some antiquarians recognize it in the large, heavy tower which rises almost from the centre of the grim castle, while others discover its ruins in the great mass of débris, of mortar and stones, which lies nearer to the town, and is often called the Devil's Drop by the common people. From both points the light could, no doubt, have been seen far away, as the cliff is high enough "to look fearfully in the confined deep," and even from the lower terraces the coast of France may on a clear day be seen distinctly.

Not one of the three ancient ligh-houses which we have mentioned can, however, for a moment be compared in magnitude and historic interest with the famous, though more than half fabulous, Colossus of Rhodes. Two thousand years ago Thucydides already complained that men received what others said about past events, even of their own country, with too great indifference, and in their indolence preferred to adopt what was thus presented without examination rather than to take the trouble of searching for truth. This experience has been amply proved by the long-credited reports about the Colossus.

Tradition has it, as is well known, that at the entrance to the port of Rhodes there was standing a gigantic statue of Apollo, with outstretched legs, one foot resting on a lower mole, and the other on a higher; holding a bow in one hand, and in the other, raised high above his head, an immense basin, in which a large fire was constantly maintained. The size of the statue, report added, was so colossal that the largest vessels could easily pass between the legs.

The facts unfortunately are, that the Colossus of Rhodes never served as a light-house, and that vessels never passed beneath it into the harbor.

The whole story rests upon the highly romantic account found in a very indifferent compiler of the seventeenth century, who for the first time mentions the Colossus as serving as a light-house, but carefully abstains from giving his authority for the statement. Another writer, of even less judgment, a translator of Philostrates, added subsequently the story of the vessels passing between the outstretched legs of the statue. This author, also, is discreetly silent as to the source from which he has derived his information.

What, then, is the truth about the Colossus? There is no lack of reliable statements concerning the statue. Strabo quotes a fragment of an epigram in Iambic verses, in which the name of the architect, Chares, from Lindos, a town on the island of Rhodes, and the dimensions of his great work, seventy yards height, are both mentioned. He adds that the Colossus was, in his day, lying on the ground, having been overthrown by a fearful earthquake, which destroyed a large portion of the city. "The Rhodians," he says, "dared not raise it again, warned by an oracle," and that is literally all the illustrious geographer seems to have learned about the Colossus. Pliny, however, gives us additional and interesting details. "The statue," he says, "fell fifty-six years after its erection; but although thrown down it is still a marvel. Few men are able with their arms to span its thumb; its fingers are larger than most of our statues. Its disjointed limbs form vast caverns, and in the inside are yet to be seen enormous masses of stone, by means of which it had been balanced. They say it cost three hundred talents—a sum which the Rhodians obtained from the sale of instruments of war, left by Demetrius before their city when he abandoned the siege in despair." A clever engineer of the third century before Christ, Philo of Byzantium, is the third author who gives, in his interesting work on the Seven Wonders of the World—if it really is his—a still more detailed description of the statue; but, as has been seen, not one of these writers speaks either of a light-house, or of the marvellous fact that ships could have sailed beneath the Colossus.

For nine hundred years the gigantic limbs remained lying near the entrance of the harbor, the pride of the inhabitants and the wonder of all travellers. In 672, however, the Arabs came, in the rapture of their first successes, to Rhodes also; and their general, one of Othman's lieutenants, caused the pieces to be cut up, and sold the metal to a Jew, who is said to have loaded nine

hundred camels with the precious burden. Thus every
trace was lost of the far-famed statue, and even the name
of the artist was long lost, although " he had made a
god like unto a god, and given a second sun to the
world."

Far different from these works of antiquity are, of
course, the light-houses of our day, in which modern
science has achieved some of its most brilliant triumphs.
England stands naturally foremost in the number of such
buildings and their mechanical perfection; for they are
of the utmost importance to her vast shipping interests,
upon which her great prosperity is mainly resting. It is
in England also that, first of all European countries, the
building and manner of lighting these towers were made
a matter of grave state interest. The care for light-
houses is there intrusted to a separate board in each of
the three great kingdoms, among which, however, the
Corporation of Trinity House, which controls those on
the English coast, is naturally by far the most important.
Unfortunately little is known as to the early history of
this remarkable body, since a disastrous fire in 1714 destroyed the larger part of its archives. We only know
that it owed its existence to a charter granted by Henry
VIII, in which it is called the Brotherhood of the Trinity House of Deptford, of Strand, and St. Clement. The
document begins with the quaint words : " According
to the sincere and perfect love and like devotion which
we bear the most glorious and invisible Holy Trinity,

and also Saint Clement the Confessor, his Majesty grants
and gives license for the establishment of a corporation
or perpetual brotherhood, to certain subjects of his and
to their associates, men and women."

Originally the sole duty of these members, men and
women, seems to have been confined to the saying of
prayers for the souls of drowned seamen, and for the
lives of those who go down the great deep.   Soon, however, more practical services were rendered by the Corporation, as appears from numerous successive charters
granted by later sovereigns.   The members were gradually intrusted with a general superintendence over all
mercantile vessels, and, to a certain extent, even over the
royal fleet.

The people had, however, anticipated their action in
erecting light-houses, and long before the Corporation
took the matter in hand beacons had been lighted all
along the coast, and were growing in number as if by
magic.   Not that the English of those days were so
wondrously solicitous for the lives of their seafaring
brethren, or so peculiarly zealous in the love of their
neighbors.   The erection of a light-house entitled them,
by a provision of ancient laws, to the right of levying a
heavy duty from all vessels who passed by the dangerous place and profited by the light they had provided.
Fortunately in this case the interest of the few became
the advantage of the many, and although after James I.
the crown claimed the exclusive right of erecting light-

12

houses and collecting taxes for their support, the number has never been seriously diminished. For the sovereigns found it as profitable as it was wise to grant or sell the monopoly to private individuals, and soon there was not a bare rock or hidden reef which was not laid hold of by some speculator in order to build on it a light-house and collect the dues. Thus Lord Grenville could find it necessary to make this entry in his note-book: "To watch the moment when the king is in good-humor, to ask him for a light-house!"

Unfortunately the system worked badly. Some of the fires were insufficient, others were neglected for weeks and months, and in all cases the duties levied on vessels were out of all proportion to the expenses actually incurred. This led finally, under William IV., to measures which resulted in a grant to Trinity House of all the royal light-houses, and of the right to purchase those that belonged to private individuals. Fortunately the Corporation was rich and could afford the heavy outlay required, especially as they continued to raise heavy tolls from all vessels.

The Corporation is nowadays divided into two classes, of which one, the Younger Brethren, numbering 361, are virtually excluded from all practical participation in the business of the society, while the Elder Brethren, 31 in number, with the exception of eleven honorary members, are the true managers of the whole department. These twenty working men are chosen from the great body of the Younger Brethren; they must have served at least four years as captains in command of large vessels, and pay a small entrance-fee upon their admission. To them is intrusted the whole care of keeping the coasts well provided with light-houses; besides which they examine and license pilots, watch over the navigation of the Thames, establish and maintain all sea-marks, admit the pupils of Christ Hospital who enter the navy, collect the revenues of the Corporation, and provide for the pensioners in their numerous asylums. The corresponding boards in Scotland and Ireland are simpler in their nature, and more or less under the control of Trinity House. England has one light-house for every fourteen sea miles, Scotland one for every thirty-nine, and Ireland one for every thirty-four.

France, formerly far behind Great Britain in the number and the character of her light-houses, has of late made such rapid progress in this direction that she has now one for every twelve miles, and these are nearly all of superior construction. The number of light-houses, which in 1819 amounted only to ten, has since risen to two hundred and twenty-four, and, like our own government, the French government also makes no charge for light-houses, but considers the duty of preventing misfortunes on the coast not as a branch of public revenue, but as a work of humanity. The whole department is under the direction of the three Ministers of Agriculture, Commerce, and Public Works, and represented by a board consisting of

navy officers, engineers, members of the Institute, and
other persons renowned in the sciences which bear upon
navigation.   Of our own Light-house Board little need
be said here, as, like the Coast Survey, it is one of the
most distinguished branches of our administration, and
looked upon abroad as one of the best models, which has
been copied, as far as the difference in the form of government would admit, in several foreign countries.

The light-houses of our day are as varied in their form
and nature as those of antiquity were simple and uniform.   Each one of them has, as it were, its own language,
in which it addresses itself to the anxious sailor.   One
bids him welcome to a safe harbor; another warns him
against a hidden reef.   This tall tower sends its light to
a distance of twenty-seven miles (60 to the degree); that
small one can only be seen within a circle of five miles.
One has a fixed light, and shines forever like a beautiful
star; another, more mysterious, suddenly blazes forth
from utter darkness, casts its welcome light far into the
distance, and vanishes as unexpectedly, only, however,
to reappear a few moments afterward, brighter than ever,
on the horizon.   Nor have all of them the same color.
Some are red, others white, green, or blue.

In spite of this great variety among them all, there is
nevertheless a general principle which governs their dis-
tribution.   Almost all more enlightened governments
have found it necessary to surround the coast with a
triple circle of lights.   The first of these consists of light-

houses of the largest class, and simply serves to define
the outline of the main land, so that the sailor arriving
from the high seas may at once be made aware of the
vicinity of land, and thus be enabled to avoid the dangers
which thicken as he comes nearer in shore. Hence all
the great capes which stretch out more or less far into
the ocean, the low islands, reefs or sunken rocks which
threaten vessels with destruction, are chosen, and on these
promontories or rocks light-houses are erected at such
easy distances from each other that no vessel can well
approach the land, unless it be in a thick fog, without
seeing one or the other. When the first circle is passed
the sailor encounters a second and a third range of lights,
of inferior size and shorter range, which warn him against
smaller reefs or sand-bars, or point out to him the often
very narrow entrance to the harbor into which he wishes
to enter. Thus the Thames is literally brilliantly lighted
up from Gravesend to the London docks, and the mouth
of the Gironde presents not less than thirteen light-houses
of the three classes. Finally, when the vessel has actually entered into the narrow channel, a fourth class of
still smaller beacon-lights greets it and guides it safely
to its precise landing-place in the " desired haven."

Difficulties, however, seemed to multiply in proportion
to the increased number of lights, for it became more
and more troublesome to distinguish one from the other;
and yet a slight mistake of this kind might lead, and
often did lead, to total destruction. Formerly this dan-

ger was guarded against by having three kinds of light —fixed lights burning steadily, and revolving lights with intervals of a minute or half a minute. But this difference soon ceased to be of practical value, partly because the number of light-houses increased too rapidly, and partly because merchantmen were too careless in noticing the small distinction in point of time. It was in this embarrassment that Fresnel came to the rescue, and made his name famous by the great improvements he introduced. Now there are seven different kinds of light in use : permanent lights; lights with a blaze, which show alternatively five blazes and five eclipses, or more, in a minute ; varied lights, which show a fixed light succeeded by a white or red blaze at intervals varying from one and two to three or four minutes; revolving lights, intermittent, alternating, and scintillating lights. Revolving lights increase gradually until they show to their full power, and then diminish into utter darkness, after which they grow once more in brilliancy, and thus they continue at regular intervals. In intermittent lights, on the other hand, the light appears all of a sudden out of perfect darkness, and disappears as suddenly again. The alternating light appears first white and then red, without any pause between the changes ; while scintillating light, the most recent invention, and due to French engineers, appears and disappears by seconds, and thus produces upon the eye the peculiar effects from which it derives its name.

There is, of course, no small skill and judgment required, first to determine the most important, prominent points which have to be lighted up in order to prevent vessels from incurring great risks ; and when that is decided, to vary their lights in such a manner as to avoid one being taken for the other. The number of light-houses along a coast is necessarily limited, not by any regard to expense, as the saving of human lives cannot be estimated in money, but by the difficulties which would arise if they were so crowded as to present to the eye, at a distance, nothing but one confused line of beacon-lights. By reducing the number to the lowest possible demand, and by skilfully varying the appearance of the light in each, distinction is made both easy and sure ; and in accomplishing this more ingenuity and labor is displayed than is commonly suspected.

Thus when the light-house has been properly placed, the great question arises, how it is to be lighted. The ancients, we have seen, had very simple means—a mass of burning wood on top of a large tower. Their light-houses were magnificent structures—beautiful monuments of their skill and their energy ; but they accomplished little. The Middle Ages made hardly any progress in the whole question of light, and it was reserved to comparatively recent days to replace in our houses sorry tallow-candles by bright and cheap gas, and the costly masses of wood or coal on the open summit of light-houses by artificial lights covered in with skilfully constructed glass lanterns. How imperfect the methods even of the last century still were may be judged of by the fact that the light-house of Cordouan, at the mouth of the Gironde, and the finest in France, diffused with its eighty lamps, each burning before a reflector of highly polished metal, so feeble a light that the mariners in 1782 unanimously petitioned for a return to the barbarous system of former centuries. The trouble was, that the lamps were half the time in the condition of those of the Foolish Virgins, and even when well provided with oil their flat wicks gave but little light, but, on the other hand, an immense quantity of smoke. It was then that Dr. Argand, a distinguished physician, who had given much time and labor to the question of light, invented the burner still known by his name. A cylindrical wick, inclosed in a chimney of like shape, with a double current of air, gave all of a sudden a light such as had never been seen before. The original system was soon perfected in many detai's, and Carcel, especially, added the method of overfeeding the lamps with oil, by which the combustion was hastened, and the vitality of the wick very largely extended. Next came the turn of the reflectors, which received not only a better shape, but also a higher polish, and increased their efficacy a hundredfold, when a method was invented by which they could be kept constantly moving around the lamp, and thus project the rays in every direction. This beautiful invention, first employed in the obscure port of Marstrand, in Sweden, was simultaneously published in France by Teulère, and at once very generally applied to all light-houses. The larger number of European maritime powers adopted it eagerly, and until within a few years it was the only one used on the coasts of Great Britain. In France this so-called " catoptric " apparatus is less generally employed, and almost entirely confined to narrow passes or specific purposes.

Science, however, is as apt and as quick to find out defects in new methods as she is slow to admit them at their first appearance. It was soon found that these beautiful mirrors not only were easily and speedily tarnished by the corrosive influence of the sea air, but that they also absorbed and thus exhausted a large portion of the light which they ought to have reflected. A Commission was appointed by the French government for the special purpose of suggesting a remedy for these defects, and fortunately the right man presented himself at once, who possessed all the requisite knowledge and genius. This was Augustin Fresnel, who had distinguished himself from childhood up by his successful studies of the great question of light, and had earned the admiration of the great Arago, whom he aided as his secretary. He discovered the lenses now employed in the so-called dioptric apparatus, and found in an optician of high merit, who bore the significant name of Soleil (Sun), an efficient assistant for all the practical purposes of his invention. The man of the great mind

12*

and the man of the skilful hand put·their united powers to the task, and the result was one of the most beautiful contrivances ever achieved. It is true that a name of great renown, that of Brewster, is mentioned in serious competition with that of Fresnel, the English claiming both the priority of invention and the superiority of construction for their own countryman; but, fortunately, the merits of both these illustrious men are great enough to be appreciated by all the world, even independently of the question connected with light-house lenses. Moreover, the improvements patented by Thomas Stevenson, the great engineer and builder of light-houses, a few years ago, and generally adopted by the British authorities, under the name of the holophotal apparatus, surpass all that Fresnel and Brewster have ever accomplished, so far as to threaten their names with comparative oblivion.

The lenses can, however, be truly efficient only when the light which they reflect is strong and steady. It has been well said that as the light which shines in front of the building is the soul of the light-house, so the lamp is the soul of the apparatus. Men like Arago and Fresnel did not overlook its importance, and introduced here also great improvements. The different light-houses are, however, provided with different lamps also. One may have a Carcel lamp, in which the wick is regularly provided with oil by means of a clock-work in the lower part; another one may be content with a moderator lamp, where a heavy weight, moving a wheel, produces the same effect; still others, of more moderate pretensions, preserve to this day the old lamp with an oil vessel on the same level as the wick. The famous savant, Rumford, who has given his name to so many inventions, from a soup to a chimney, first suggested the idea of increasing the illuminating power of common lamps by providing the burner with several concentric wicks; but he was unable to carry out his plan. Arago and Fresnel fell heirs to his idea, and by dint of hard labor succeeded at last in carrying it out; it is to them we owe the present improved lamp, which gives out an intensely strong white light, and yet continues to work for more than twelve hours without requiring any attention. The great advantage of this feature can only be fully appreciated when we bear in mind that these lamps have to remain burning during the whole time of the longest winter nights. Since this system was inaugurated, lighthouses of the third class have lamps with two concentric wicks'; those of the second class have three, and the largest even four such wicks. It would be an error to imagine that the flame itself is larger than ordinarily, although the larger apparatus produces a light equal to that of twenty-three Carcel lamps; the flame is only moderate, but of perfect whiteness and dazzling intensity.

The material used for illuminating purposes is oil of various kinds in England and our own country; in

France rape-seed oil is almost exclusively used, though petroleum is beginning to supersede it in many districts. In one instance only an electric light has been tried; it is far superior to all others in brilliancy; but the expense is serious, and the danger connected with the process of production so great as to make it as yet unprofitable and inexpedient. There is no doubt, however, that it will ere long be made both cheaper and safer, as the electric light used on board large vessels, and first introduced by Prince Napoleon, has already proved itself of the very greatest usefulness.

If the light is the soul of the light-house, and as such all-important, it has, after all, to be clothed in a body, and the house requires for its part hardly less consideration. It has been shown that in ancient times the form was apt to be more or less fantastic; in our day it matters little whether the tower rises on a lofty promontory overlooking land and sea, or on an isolated rock surrounded by turbulent waters; its construction is subject to certain laws and rules which the engineer dare not neglect. He must provide for them a suitable form, great strength and stability, and perfection in all details.

The height of light-houses varies, of course, according to the place which they occupy; as a rule it has to be very great in order to enable mariners to perceive the friendly light from afar. Hence they are not unfrequently placed on top of a mountain or the summit of a cliff, and then the tower need only be sufficiently high to rise with its lantern above trees and buildings, and to be secure from wanton injury and the contact with small stones raised by the tempest. If, on the other hand, the light-house must necessarily be on the coast, or even out in the open sea on rocks nearly level with the surface, then towers are required of at least 120 feet, and there are structures of this kind on the English and French coasts which exceed even 200 feet.

The towers are now almost always cylindrical, and of small diameter. On land they are surrounded by buildings intended for the keeper and his family, the visiting engineer, and at times even for farming purposes. Those out at sea present, of course, a very different appearance. A strange ladder affords the only means of access. It consists of strong bars of copper let into the rock of which it is built, and carefully cemented. As we ascend we come to folding-doors of bronze, heavy enough to require the full strength of a man to move them, and hermetically closed so as to protect the entrance against the heaviest swell. A long narrow passage, looking as if it were cut in the live rock, receives us as we enter the lower story of the light-house. Here are large quantities of wood, ropes, and timber stowed away. A story higher we see the enormous tanks of zinc in which the oil is kept which feeds the lamp above, and the water on which the life of the inmates depends. In the third story is the kitchen and a store-room, on a level with the first gallery which runs around the tower. We pass the doors of

three small rooms in which the keepers live, and continue
to ascend till we come to the seventh story, where we are
invited to rest for a moment in a snug little parlor of octagonal shape. This is the room reserved for the engineer,
who comes from time to time to inspect the light-house.
It is comfortably furnished, and displays in the arrangement of the furniture, which includes a large bedstead, all
the ingenuity familiar to men on board large vessels.
A few more steps on the spiral staircase lead us to the
more important parts of the tower. The eighth story
contains vessels for oil, spare lenses, reserve lamps, and a
few delicate instruments for meteorological observations.
Here the staircase ends, and we see a low vaulted ceiling
supported on a slight pillar. A slender ladder of cast-
iron leads us into the room in which every night one of
the keepers is on watch. It is strangely ornamented
with slabs of marble of various colors, which cover the
ceiling, the walls, and even the floor. We are told, in
explanation of this apparent extravagance, that it is the
result of necessity; for the illuminating apparatus hangs
down from above into this room through a circular opening in the ceiling, and makes it necessary that the room
should be kept in a state of scrupulous tidiness. This it
has been found can only be permanently obtained by lining it throughout with highly polished surfaces. We ascend one more flight of steps, the tenth and last, and find
ourselves in the cupola itself, in the centre of which is
suspended one of those marvellous lamps which make the

boast and the glory of modern science. The room is inclosed in a huge lantern of glass, and covered with a dome of copper, surmounted by a lightning-rod. The panes are extremely thick, and yet they are not unfrequently too weak to resist the wings of sea-fowl, whom the brilliancy of the light attracts. Even land-birds, traversing the sea by night, are occasionally overtaken by hard weather and dashed against the rigging of ships at sea or the sides of light-houses, and in the morning found dead on deck or among the rocks. It has been conjectured that, disliking the uproar in which they are enveloped by the storm, they make voluntarily toward the strong beacon-light in search of an asylum; but it is quite as probable that amidst the fury of the winds they lose the power of directing their own flight, and are dashed accidentally against the lofty tower. It was by flocks of such distracted birds that once all the nine windows of a massive lantern were broken in the same night; and in another instance a wild-goose, after having broken a pane, flew in between the costly mirrors and fell into the flame itself, finding there a miserable death. How numerous these strange visitors are may be seen from the fact that one thousand sea-fowl were once taken in one night by the crew of a floating light-house, and converted by them into a goodly number of gigantic pies. Fortunately not all sea-birds are equally dangerous, and in one instance at least they have actually been taught to render mankind an eminent service. There is a superb

light-house on the South Stack, a huge rock connected
with Holyhead by a suspension bridge, and abounding
with sea-fowl, who build their nests in countless caves.
These gulls settle in flocks on the walls of the light-house,
and warn by their piercing cries the mariners who might
approach within a dangerous distance.   Formerly the
tower was provided for this purpose with a gun and a
large bell, but the natural guardians were found to be so
much more efficient that the cannon was removed to some
distance, in order not to disturb and frighten the birds.
On the rock itself the young gulls are seen playing with
the white rabbits, who seem to look upon them as merry
companions, and both are most pleasant society for the
lonely keepers, shut up as they are in their tower, against
which the winds and the waves are continually trying
their strength.

For in speaking of the soul and the body of these
light-houses we must not forget the poor fellows who
are shut up within, and who often have not only their
little joys and their long sufferings, but even their startling adventures, which have more than once furnished
the material for soul-stirring recitals.

It is the custom of almost all countries possessed of a
navy and large mercantile fleets to take the keepers
from the vast number of disabled seamen.   There are
generally three of them in larger light-houses, and as a
matter of course never less than two, even in the smallest.   Their duty is simple, but exceedingly rigorous.   It

matters not whether their lonely home rise from the
waters of the ocean, miles and miles from every human
habitation, or whether it stand at the entrance to a
large harbor, crowded at all seasons with a host of vessels; the waves may dash furiously against the sides of
the tower, and try to send their spray into the very lantern above, or they may never do more than gently kiss
the foot of the building; the sea around may be crowded
with ships of all sizes, from the vast ocean steamer to the
lumbering sloop, or the eye of the keeper may be strained
in vain to perceive even the low sail of a fishing smack—
as soon as the sun sets he must light his lamp, and as
day breaks he must put it out again. During the day
his time is taken up in cleaning the apparatus and making all ready for the hour of darkness. In the larger
light-houses, where the number of keepers is greater,
they can enjoy each other's company, have their houses
and little gardens, and a certain amount of liberty, and
even enjoyment. In the smaller towers, on the contrary,
where the position is such as to make the erection of
buildings impossible, and where only two keepers, or
even a single one with his family, are left to their own
resources, life is necessarily sad and monotonous, almost
beyond endurance. In summer they may amuse themselves by fishing; and ingenuity has taught them a
novel method, invented by some unlucky man, whose
tower stood so completely isolated amidst the waters
that he had not standing-place enough on the rocks to

cast a line. This led him to fasten a line at a certain height, but just below the entrance door, around the whole tower, and to this line he tied fifty or more smaller lines with baited hooks. When the tide rose the fish were seen swimming all around the tower, they were tempted by the bait and hooked, and when the tide fell there was seen hanging around the light-house a rich garland of all kinds of fish. On the other hand, the poor keepers have their trials also. At times the wind blows with such force that they are hardly able to breathe, or the weather is so bad that they are obliged to keep the tower hermetically shut for days and days, and see from their darkened cell nothing but impenetrable fog without, and the foaming crested waves like dim and dismal shadows. It is but rarely that light-houses become the scenes of great and startling events, as was the case in the well-known instance when brave Grace Darling rescued the shipwrecked passengers of the steamer "Forfarshire," and made her name a household word with all who admire heroic devotion and true Christian courage.

In spite of the monotony of this life, it is yet not without its admirers, and Mr. Smeaton tells us of a shoe-maker who applied for a place as keeper in the Eddy-stone Light-house, because he was tired of the loneliness of his shop! He found that he was less alone on his rock than he had been in his narrow alley, and replied to those who expressed their astonishment at his choice:

"Every body has his taste, and I have always liked independence!"

Another keeper, at the same forlorn place, seems to have had by nature a clear vocation for his profession; at least he had conceived such an attachment for his strange home that he would never leave it, and even refused the short leave of absence to which he was entitled every year. At last he was prevailed upon to give the outside world one more trial; but he had no sooner mingled with other men than he felt quite forlorn; he lost his self-control, and after having been for long years the most regular and correct keeper of a light-house, he suddenly became a drunkard, and committed all kinds of excesses. He had to be carried back to his tower, where he died after a few days' sickness and suffering. Others, on the contrary, have been attacked with insanity from constantly beholding the same scenes and receiving for years the same impressions. At the distance of about a mile and a quarter from Land's End, and on a group of granite rocks surrounded on all sides by water, there rises an old tower called Longship's Light-house. It is built upon a rock of conic shape, which raises its narrow summit nearly forty-five feet above the level of the sea. In the winter the waves often rise to such an enormous height on this stormy coast that they completely hide the rock and the tower for a few seconds behind an impenetrable veil of foam and spray, and not unfrequently great injury is done to the building. Thus the sea once carried away the top of the lantern bodily, entered into the tower, extinguished the lamps, and could only be mastered by great exertions and remarkable presence of mind. Another circumstance contributes not a little to the horror of the place. Under the rock on which the tower stands the waters have washed out a deep cavern, which communicates by a narrow crevice with the open sea; when the weather is stormy and the waves are high the compressed air in the cavern produces such a fearful roaring that the men cannot sleep, and a new-comer was once so terribly frightened by the unexpected noise that his hair turned gray in a single night. Six years ago the people on shore noticed two black flags fluttering from the flag-staff of the light-house. They surmised at once that some great calamity must have occurred. A boat tried to go over, but the weather was so bad that the brave men who ventured their lives in order to reply to the sad signal of distress, had to wait for some time, and then only reached the rock at imminent peril. The scene which presented itself to their eyes was horrible. One of the three men who lived in the light-house had, when his turn came to go on duty, in a fit of despair cut open his breast with a large knife. His companions had endeavored to staunch the blood by stuffing pieces of tow into the wound, and three days had passed without their being able to obtain assistance. Even now the sea was so rough, and the difficulty of getting into the boat

so great, that the wounded man had to be let down in a
rope chair. In spite of every care and attention the
poor fellow died a few days afterward, and the jury rendered a verdict ascribing his death to temporary insanity, caused by his long-continued isolation on the
rock.

Nor is it the least of the evils connected with light-
houses that frequently men of most uncongenial temper
are forced into unbroken intimacy by their common
imprisonment on a lonely rock. A curious traveller who
visited the famous Eddystone Light-house once asked
one of the keepers if he was not, after all, quite happy in
his tower? "Oh yes," replied the keeper, " we might be
very happy here if we could only have a chat with each
other. But here is my chum, he and I have not exchanged
a word with each other for a month ! "

Such are some of the marvels connected with light-
houses, and such some of the features of their inner life.
We may hereafter endeavor to state what noble efforts
have been made in our day to improve these interesting
structures and to surround them with varied auxiliaries.
For, to the honor of our age be it said, the nations most
interested in the subject — England, France, and the
Union—have all well understood the duty resting upon
them in this respect, and given to it all the attention it
deserved.

For surely no expense ought to be spared, no amount
of mental labor counted lost, which may contribute to

the perfection of light-houses, when we remember that on
the coasts of Great Britain alone, in a single year, nearly
one thousand vessels were wrecked, of which half were
totally lost, the rest stranded and seriously damaged.
But the loss of treasure was insignificant in comparison
with the far greater loss of one thousand and five hundred
lives! Surely, then, nothing ought to be omitted by a
great maritime people that could reduce the annual loss
of lives, and thus render good service to humanity; and
leaving out of consideration for the present the admirable life-boats of our day, no other means are more efficient for this noble purpose than the erection and judi-
cious management of well-appointed light-houses.

X.

A GRAIN OF SAND.

"As children gathering pebbles on the shore."—Milton.

IT is a common error to suppose that only the lives of the great and the noble are full of exciting adventures and remarkable incidents. Even among men the lowliest often experience the strangest fate, and the eventful career of the English blacksmith's son, who ended his days as a great commander, and lies by the side of a Dante in a Florentine church, is far more attractive than the life of many a renowned general or brilliant statesman. The same is true in nature. The king of animals grows and mates and dies in his desert lair, and the battles he has fought with grim rivals, and the number of sheep he has stolen, or antelopes which he has killed, are all that can be told of the terrible lion. The little bird, on the other hand, that flew frightened and fluttering to the topmost branch of the Mimosa, when his deep roar made the earth tremble for miles, may be carried a captive over wide seas to distant lands, and pass through scenes and trials which would fill volumes and excite our wonder. The lofty mountain stands for ages unmoved on its firm foundation, and generation after generation passes away at its foot without noticing a change in its form or nature. But the little grain of sand, which the waters of a mighty river roll sportively toward the great sea has a history full of events, and might tell us, if we could hear its low voice, a tale of strange import. We will try to follow it from its birth to the place where we saw it first—on the banks of the Rhine. Like one of the fabled heroes of old, its birth-place is wrapt in mysterious doubt. Some say that Neptune sent it forth into the world and bade the obedient waves to give it form and shape, while others insist that Vulcan fashioned it in a fiery furnace. Perhaps both divinities helped to create it, for we find it in the shape of a quartz crystal of goodly size, lying snugly ensconced in the heart of a sharp-edged cliff of gneiss, and gneiss is said to owe its origin to the repeated action of fire and water in rapid succession.

For thousands of years the crag has been standing on high, reaching boldly up into the blue ether, and crowning an Alpine mountain with its clear outline, distinctly visible from afar in the pure, transparent air of mountain regions. Many an avalanche has come down from still greater heights, and passed harmless to the right or the left of the sharp tooth, or, falling headlong upon the pointed rock, has broken its force there, and filled the air far and near with clouds of powdered snow. But at

last one greater than all before came down from the
highest of those inexhaustible deserts. For years the
white, treacherous field had moved slowly and imperceptibly toward the precipice, until it was cunningly
poised on the edge. A sudden current of air, the foot-
fall of a child below, or the merry song of a shepherdess
broke the charm that held it suspended on high, and in
an instant the whole mass, but just now so silent and
motionless, leaps forth with the swiftness of lightning and
wakes the echoes for miles and miles with its terrible
thunder. As if endowed with a very rapture of motion,
the huge white mountain rushes forward, darkens the
air with its portentous shadow, and aiming straight at
the bold crag, wrests it with irresistible force from its
ancient foundation. Other crags and cliffs that had
once formed part of the great Alpine mountains, and
protected them against the terrible enemy, had probably
fallen one by one in fruitless struggle, and now the last
of old landmarks was to be effaced.

A few seconds more and the sky is serene as before,
the scene as silent as ever. Now and then only a little
white cloud rises lightly in the clear air, as a small mass
of snow that had been caught by a tree or a jutting rock
crumbles to pieces, or a low, rumbling noise tells of a
block of stone that rolls downward after having halted
awhile against a gigantic pine-tree or a rock by the way-
side. Soon, however, this also ceases, and all around is
once more still as the grave.

13

The old crag still rises heavenward, and stands out black and bold amid the desert of fresh fallen snow that surrounds it on all sides. But its crown is gone; huge masses of rock have been broken off and are buried far below, and among them is hid the quartz crystal with our little grain of sand.

Loosened and driven away from its ancient birthplace, it lies buried beneath towering masses of snow in the bed of an Alpine torrent, that comes raging and roaring from the higher hills. The avalanche has laid itself insolently across its bed, and the poor little rivulet frets and foams as it tries to pursue its course toward the valley beneath. But in vain it attempts to scale the half-frozen mass and to send its water over the back of the giant; it has to learn patience, and slowly begins to overcome the oppressor, not by might, but by perseverance. By day and by night, in sunshine and under dark clouds, it gnaws at the foot of the gigantic walls till an opening is made, and it runs once more merrily under the high arches of a snow-built bridge. As spring changes into summer its fury subsides; the foam clears off, the roar is subdued, and the brook resumes its wonderful rise and fall, being full to overflowing in the early afternoon, and low and silent in the morning, as the melting and the nightly rest of the glacier command it, from which it draws its unceasing supplies. At the very bottom of the rock-covered bed of the torrent lies the block of gneiss that holds our little grain of sand, and busily does the water play around it, year after year, washing it with its milk-white waters, that are full of the glacier sand, and continually beating it with larger pieces of stone which it brings down from on high, and rubs and pushes against the patient wanderer.

Years pass by, and the stone, growing smaller by degrees, makes slow and almost imperceptible progress toward the plains, till at last it finds itself lying at the edge of a rocky precipice, over which the brook sends its waters in savage joy. They fall between the perpendicular walls, which they have smoothed and polished with incessant labor, down into a dark abyss, from which they rise immediately again as airy vapors, reflecting the rays of the sun in ever-varying colors, and every now and then mocking the rainbow on high in all its gorgeous hues. One fine day a warm wind, coming all the way from the dread desert of Africa, fills the torrent with floods of molten snow and ice from the glacier above; it begins to rage as of old, and in fiery tumult it carries along with it all that fills its narrow bed; trees and stones and débris of every kind must follow its mad career; and as the swollen waters throw themselves headlong in a huge arch from the edge into the boiling turmoil below, our block also is forced to take the fearful leap, and falling upon a long-tried rock of unwonted hardness at the foot of the cataract, it breaks into numerous pieces.

The good results of the old policy, *Divide et impera*,

are immediately visible; the stone is reduced to the size of a man's hand, and now rolls readily enough before the pressing, pushing waves that continue to come tumbling from on high. Further and further it travels from its first home, and at every step and every leap it knocks against its travelling companions, taking from their substance and losing more and more of its own, till the parts that have been united for centuries are divided forever, and each has to go on its own path through the world.

Our little grain of sand, however, is not yet independent; it still lies concealed in its diminished rocky home, and has silently and patiently to await the hours when it also shall have to fight the battle of life by itself.

All of a sudden, however, it is exposed to a new and terrible danger. After its long quiet rest in its airy home, and the slow and measured progress it has been making of late, as the waters of the brook roll it leisurely along, comes unexpectedly a time of fearful excitement. On one side of the brook lies a huge block of granite, almost level on the top, and so high that the waters play over it only when the river is at its fullest. The stones and pebbles which they carry have gradually worn a slight indentation on the surface, and now every stone is forced to turn round and round in the little hole, until the water subsides once more. The dance looks merry enough to the beholder, but it is full of sadness for the performer, who is ground down day by day by the incessant friction, and must perish, being reduced to powder, unless a sudden rise or a lucky accident releases him from the bondage and sends him back into the bed of the torrent.

Our block also has been forced to dance madly around, until it has lost more than half of its size; but at last the brook is reduced, the heat of the sun dries up the water in the little pan, and the pebble lies weary and exhausted amid the dust and dead insects which line the bottom. But ere long the torrent resumes its mad career; the rock is covered with water once more, and the furious dance recommences. Thus the poor little grain of sand alternates between perfect peace and restless excitement, wearing away steadily all the time, till at last a fellow-sufferer comes, and pushing it over the edge, takes its place in the strange ball-room.

The grain, still hid in its cradle, now continues its slow journey. But a furious rain sets in for days and nights; all the mountain streams swell and rise, till their bed is too narrow, and the brook, filled to overflowing, spreads its dimmed waters over a vast meadow. The pebble, born amid ice and snow, and restlessly rolled along among rocks, now for the first time is surrounded by bright flowers and waving grasses, but it is still pushed along by the restless waters of its torrent, and crushed beneath larger stones and boulders, while the brook has been borne on high, when it aped the great river, and now slowly rolls in the icy lava-like mass of débris over the blooming Alpine meadow.

Behold the little grain of sand now on its new stage of life, shrunk and shriveled without, with corners abrased and edges dulled down, but still dwelling safely within its old stone prison. The increasing pressure from above, as the burden, borne lightly by the swollen waters, becomes too heavy for their diminished strength, and gently sinks to the ground, continues to break and crush the gneiss, till at last there remains but a stone of the size of a hazelnut, which serves as a dark, dismal dwelling for the tiny grain.

It has not long to wait for the day of liberation. A new tempest pours overwhelming floods of rain into the same deep channel, from which years ago the mountain torrent rolled its waters over the fair meadow. They rack and rend the stony bed to its foundation, and our luckless adventurer is tossed and torn about in all directions. A heavy block comes thundering down from on high, and falls with loud clatter upon the diminutive piece of gneiss; the little plates of mica give way, the outer shell falls to pieces, and for the first time in its life the grain of sand lies open to the air and beholds the bright light of heaven. It is washed by cool waters, and when the sun once more breaks forth from behind the dark clouds, its rays are reflected in playful glitter from the sides of the tiny grain.

As a person leaving a public assembly is borne along for a time by the crowd, and at last escapes from the rush and the press to pursue his own way at leisure, so the

grain of sand also at first was carried onward in a new,
narrow channel, which the swollen brook had furrowed
out across a meadow, staying now in this bend and now
in that for a while. Finally, however, the little branch
falls into the main channel once more, and the still im-
petuous waters carry all its contents with a sudden splash
and splutter into a small Alpine lake. There the little
grain of sand sinks at once to the bottom, where it meets
a numerous company of other unfortunate grains, and
seems to be doomed to remain there forever, for what
power is likely to take the helpless pebble from the silent,
motionless deep? The lake has no outlet through which
the additions made by avalanches, rain, and snow could
be carried off; and yet the marvel is, what becomes of
all the vast masses of water by which its little basin is
filled to overflowing? For nine or ten months in the
year the lake is frozen, and can, therefore, lose nothing
by evaporation; and no river or outlet ever finds its way
out of the deep, circular loch in which it is confined.

It rids itself, nevertheless, of all it cannot hold, though
the eye of man cannot watch its proceedings. For the
pressure of the heavy mass of water which it contains in
a narrow basin is powerful enough to force some of it,
drop by drop, through fissures, cracks, and crevices in
the rocky bottom, and these minute rivulets carry naturally with them all that can pass through the narrow
passages, which are their only outlet. Eye of man, we
said, cannot watch this process, and yet it is not unseen,

for a slow, scarcely perceptible motion on the surface, by
which wide circles are continually formed, proves to the
experienced observer, that the whole lake is a gigantic
funnel, from which water is continually though slowly
escaping.

One of these minute trickling currents sucks up our
little grain of sand, and so bears it downward through
the dark, hidden channels of the rock. How sad seems
its fate, to be torn from its lofty home high in the free
air, and now buried far under ground in eternal night!
Enemies, moreover, surround it on all sides; now larger
stones come and try their strength against the little pebble,
and now subterranean waters meet and attack it with
the acids they contain. But its nature is strong; quartz
is hard to overcome, and, though it may lose here a little
of its rough outline, and there a particle of its substance,
the grain still preserves its identity and continues its
strange pilgrimage.

Far down in a low valley, on the nethermost terrace
of the Alps, there lies a meadow rich in beautiful flowers
and fragrant grasses. On the shady sides beneath noble
beech-trees and maples dwells a whole host of graceful
ferns and tiny mosses, while out on the sunny plain the
green carpet is thick and deep, and shines in the bright
light with velvety richness. In the centre of the sunny
slope, right amidst flowers and ferns, a powerful spring
bubbles forth. Its crystal-pure waters flow gently downward over a bed strewn with white pebbles, and never

fails in the saddest drought of summer or the bitterest
frosts of winter. Here is one of the secret outlets of the
upper lake. A few hundred feet further on, the waters
are gathered in a little basin, as if they wished to rest
after their painful creeping through dark crevices, and
gather new strength for the weary journey that is
finally to carry them to the great ocean.

Very different, however, is the miniature basin here
below from the little lake on the heights, from which it
receives its supplies. Here animals of all kinds creep
and crawl busily on the bottom, and a wreath of reeds
and aquatic plants surrounds its low, marshy banks.
Only on one side there is a break in the fringe; it is the
place where the waters of the spring come forth to continue, often a short rest, their long journey through the
valley.

Once more the little grain of sand greets here the
light of day, welcomed by a thousand fair flowers upon
its return to life. The spring has brought it forth from
its wanderings in the dark; it lies snugly ensconced
among sheltering grasses by the side of the opening, and
the rays of the sun lighting up its transparent edges
make it flash for a moment like an eye beaming with joy.

Soon it becomes a toy of the merry waters, now being
tossed on high in playful sport, and now left unnoticed
by the side of the channel, till at last it drops wearied
and worn into the little basin below. In the upper lake,
large as it was, there was no fish whose swift and easy

13*

movements ever stirred the quiet waters and threatened
to dislodge the tiny grain; there was no bird there to
seek its prey by the bank, and, in the search, to tread
upon it with its broad web-foot. But here below, in the
diminutive basin, dangers at once surround it on all
sides, and before it is well at home in its new resting-
place, it is carried once more to distant lands. A beautiful duck comes with loud whirring flight; it plunges
headlong into the water to dip up with its broad bill a
whole mouthful of worms, and then, rising to the surface
again, it enjoys for a moment the play of the sunlight on
its bright green neck, and lets the water-beads glide
harmless from its brilliant plumage. Now it flies off,
and with it the grain of sand goes to a new home. For,
as the duck gobbled up a little leech, the animal had
just sought safety by fastening itself with its suckers to
our tiny wanderer, and thus the stone was swallowed
with the food. The bird, far from complaining of the un-
welcome visitor, enjoys it hugely, for it needs the sand
for digestion, and the little grain finds there neither a
grave nor a resting-place, but is called upon to work,
and to work hard in its strange aërial ship. It has
become a diminutive millstone, and in company with
others who have preceded him in the strange mill, it has
to go round and round in its dark cell, a lifeless but
useful help to organic life.

How the grain of sand now wandered with the duck
from lake to lake, to-day through long, lonely regions in

the pure air, to-morrow through dark rushes and dismal swamps! After a few weeks it is left by the bird on the banks of a distant lake, where busy laborers are forming bricks from the red clay of the soil. All of a sudden the little grain of sand finds itself seized by human hands; it is worked and kneaded, and finally finds itself lying incorporated in a hollow tile, on a drying-frame. The warm air has soon dried up all the water that was in the clay, and the tiles and bricks are placed in long rows in the kiln. Here new horrors await the unlucky grain; fearful flames break forth all around, and hungry tongues find their way into every crack and opening, till the fiery breath enters deep into the core of the bricks. They blush with indignation, and are drawn out to be carted away to a new house on the banks of the Rhine.

A few days more and the grain of sand hangs once more high in the pure air, on the surface of a tile on the roof. The waters have reduced it in size till the careless eye would hardly observe it; the fire has cracked it, and a ray of the evening sun plays merrily on the small, sparkling eye. Once more banished and bound, it remains there at rest for many a year; little birds sing by its side in bright summer days, and in winter the snow covers it up with sheltering care.

But it is not at rest forever yet. The hidden powers of the air come to their silent work, and gnaw and nibble at the hard clay. They call in the aid of wind and weather, of rain and snow, of light and heat and frost, and ere long the hard tile is changed into fertile soil. In visible germs come on the wings of the wind, and minute mosses begin to cover it with their green velvet. They grow to tiny bolsters, and one of them covers up the grain of sand, condemning it once more to dismal darkness.

Thus it cannot see the black cloud that rises in stern threatening from the east, to summon it to a new pilgrimage. For a moment man and beast are filled with strange fear, and then the tempest breaks; a fierce storm of hail-stones falls far and near, and in the next minute the tile with its grain of sand lies broken on the ground by the side of a hundred companions.

The day after a cart comes and the useless fragments are carried to the banks of the river, where familiar voices greet the grain of sand and tell it many a tale of the world above, where the waters were born, and where its own childhood was spent. "Come along," they whisper and shout; "come with us; we'll carry you to the great ocean."

And again a fearful tempest breaks down upon the glaciers in the high Alps, and again fills to overflowing the countless rivulets that form the great Rhine. The rain comes in floods, the glaciers slide downwards, and the masses of snow on the lower peaks melt under the hot breath of the African wind. Down into the valleys they all rush, now singly and now united, till the bed of the great river cannot hold the vast abundance, and the waters rise high above the banks on both sides. With irresistible force they carry away whatever is not strong enough to cling to the soil, and thus the broken tile also is seized by the enraged giant and borne on to the falls that are not far distant. A second time the grain of sand must follow a master, who holds it in his strong grasp, in a fearful leap over a precipice, and down in the abyss the poor wanderer is again whirled round and round in the seething maelstrom, till, sadly shorn of its proportions, it is at last pushed forward, a tiny grain in truth, into the wide plains where the waters are quiet and flow on in peace.

The adventures of the little wanderer are over now. For the rest of its life the grain of sand is to be gently pushed along—who can tell how much or how little every year?—by the waters of the mighty river. It has been freed from its bondage to the fire-burnt tile, and lies now, the high-born son of a lofty Alpine peak, on the dark bottom of the river. A freshet may now and then throw it for a time on the sunny bank; and accident may bury it for years under some great rock, but, after all, it moves on steadily, though slowly, on its heaven-appointed way to the sea, and whatever may befall it, whether it be ground to dust or dissolved by acids, so much is certain —it will never be lost! For there is nothing lost in the great household of nature; and the glory of the Most High is not more strikingly displayed in the rising and setting of the great orb on high than in the varied fate and the unseen adventures of a little grain of sand.

XI

MERCURY.

"Swift, subtle Mercury."—Roger Bacon.

GOD said to man, whom he had made in his own image: "Let him have dominion over all the earth!" and man, from the day of creation to our own, has labored hard to make himself master of the world. For like all other gifts from on high, that power also has to be earned in the sweat of his face, and the ground that was cursed for Adam's sake yields no longer willing obedience to its sinful master. Nevertheless, there is no man so poor but he can control all that he sees, no race so brutal but it makes all Nature serve and work for its benefit. Even on the confines of the habitable world, where the blessed light of day shines but for a short time, and ice and snow sit a barrier to all life and enjoyment, even there man still succeeds in ruling the elements, and employing the scanty supplies that the earth can yet furnish. The low-creeping moss and tiny sweet berries preserve his health, while the grossest food gives him support and warmth; the monster of the seas gives him

bones for his arrow-tips and timber for his fragile boat;
the sea-lion furnishes oil for his lamp, and the very ocean
supplies him with a house which he rears by the aid of
huge masses of ice.

Among all the vassals, however, the goodly crowd of
metals are his most useful servants, his best friends.
Buried deep down in the bosom of the earth, hiding
themselves in darkest corners and remotest chambers,
they seemed to have fled from his all-absorbing rule.
As if anxious to avoid the slavery that awaited them in
his service, they had run into tiny veins, split into thin
leaves, scattered into minute grains, to escape his attention: here they covered their bright splendor with unsightly clay; there they mixed with worthless earths,
and often they allowed the merry waters of mountain-
streams to roll over them and make them invisible. All
in vain! For man seized upon the last of the fugitives,
whose bright faces he caught here and there on the surface; he followed them on lonely paths through the
mountains; he pursued them with spade and pickaxe into
the very bowels of the earth; he sank shafts and filled
them with monstrous machines, that forced them to come
forth from their hidden recesses, and seized them where-
ever his knowledge revealed to him their retreat. Then
he carried them triumphantly up to the surface of the
earth, beat and stamped, ground and crushed them,
threw them into furious fire, until they yielded to his
stern will, and assumed the form which it pleased him

to give them; he chilled them, and hardened them, and at last took them into his service, forcing them to do his bidding, and to aid him in becoming truly the master of all the earth. For man had dominion given him over the inert materials as well as over the living creatures of the earth. Hence the true poetry of mechanics, attractive in all its marvellous doings, and more charming in its daily results than the wildest dreams of creative fancy. Let those who will, see nothing but masses of clanking iron and huge incessant fires, nothing but tall chimneys and clouds of black smoke; to the imaginative, even smoke and the vapor we call steam becomes an embodied genie, who raises man to the clouds and at whose feet the earth opens at command; and those who yield themselves to the spell are led through subterranean ways to the secret chambers of the treasures of Nature. Or, led by the same obedient slave, they find themselves in gardens more enchanting than any Aladdin ever saw—gardens of vast extent and varied beauty, covered with transparent crystal, containing all beauteous things that Nature produces or the immortal mind of man creates, with the flowers of all zones and the fruits of every land, with living marvels all around, and fountains throwing out liquid gems, with a night as dazzling as the days are brilliant. And this is the romance of reality.

For man has dominion over all the earth for some good purpose; he is the steward only, and, far from indulging in wanton destruction, he makes his vassals his faithful

servants, his attached friends. Even the humblest of
Nature's children becomes useful, when he names it, and
by his heaven-born instinct assigns it its duty. The air
has to fill his bellows, and the fire to work like a slave by
night and by day; the water must fertilize his gardens,
and the stones afford him dry paths ; the falcon hunts
for him, the fir-tree carries him across the ocean, the very
worm serves him as bait, and the lowly herb restores him
to health.

But of all his servants, the metals, those kings of the
old alchemists, which were so nearly related to the great
heavenly bodies that they bore their names, have ever
been his most useful vassals, his best friends. As he
learnt to know them and to employ them, they changed
his whole manner of life, and he counts his own history
from the age of bronze and of iron. He subjected them,
one by one, to the manifold evolutions of the great workshop in his brain, and gave to each a life of its own.

Thus he saw at a glance the stubborn strength and the
enduring power of iron, and called it up from its dark
hiding-places to become his workman, to subject the
whole world to him by its strength, and to embellish it
by its numberless uses. He summoned it to check and
control the beasts of the earth, and iron changed into a
supple, cunningly-woven chain to fetter the wild bull;
it became a bit and a spur to master the proud, prancing
horse, a sharp-pointed staff to strike even the huge elephant with terror, a solid cage to hold the lion, the king

of the beasts, and a slender hook to catch the fish in his subtle element. Man ordered it to conquer the earth itself, and as plough it drew cruel furrow in its bosom to bear abundant fruit for his support; it became an axe to fell the loftiest trees, a saw to divide the gigantic trunks; and then it assumed a thousand varied forms to build him his house, to hold it together against wind and weather, and to protect it with lock and latch. The very Proteus of the metals, it took a new shape, and conquered even the hard rock, smoothed it and shaped it into beauteous forms, or piled it up higher and higher in majestic temples and lofty cathedrals; or it took the tallest of pines on high mountain-chains and bound them with clamp and clasp to form a vessel, and held it as anchor to the bottom of the mighty deep. Soon there was no work done in kitchen or parlor, in workshop or laboratory, in which iron did not show itself an ever-ready, ever-handy assistant, yielding with willing obedience to man's will, and obediently assuming every shape he desired.

Man rejoiced in the skilful laborer, and devised new and harder tasks; he lent the whole power of his mind to make new inventions, which iron had to help him in carrying out, and, ever hand in hand, the master and the servant went onward on their path of improvement. The more docile the metal showed itself, the more rapidly man's progress made itself felt on the earth; and with the labor he performed by the aid of iron, his own spiritual power increased by degrees. Whatever had appeared impossible before, iron had to achieve. Long had the great streams of the earth impeded the commerce of men; for they defied the structures of fragile wood, and of massive stone, when high floods gave them unusual power, or huge drifting blocks of ice made playthings of pier and bridge. At last, here also iron came to the aid of man, and once more obeying his ingenious command, it stretched out into long rods and slender beams, chained them one to the other, and lo and behold! the gossamer chain-bridge hung high and safe above the furious waters, and man could defy now, on his part, the power of the conquered element. And when he had thus overcome the last obstacle that impeded his free control of space, he became impatient of the swiftness of the horse even, and bending his mind to devise some more rapid mode of overcoming space, he invented a new service for his trusty servant. Iron was tortured and twisted anew, until it assumed the form of colossal wheels, huge levers, caldron, and pipes—and the engine was ready to bring with its own amazing uses the power of steam into the service of man. The new servant became the most powerful and the most delicate of instruments; here moving machinery of gigantic proportions, and there printing the tiniest characters on silk and on cotton. It carried man with amazing rapidity from land to land, from continent to continent; and as the cunning web of the spider holds the strongest of her enemies

helpless in its meshes, so the iron net of railroads and
steamers held the very earth captive at the will of man.

Even greater things, however, he demanded of the
faithful metal, which he had discovered possessed a marvellous gift appearing in many ways not inferior even to
the instinct of living creatures. You must show me the
way, he told the willing servant, across the pathless
ocean! and iron changed into a small needle, and as
compass and sextant it became his unfailing guide over
the broad ocean and around the whole globe. But when
man rose against man, and fierce war raged through the
land, even then he bethought himself of his faithful friend,
and iron came to his aid, now as a sharp sword, and now
as a gun or a cannon. Thus, in peace and in war, on
land and at sea, the useful metal is by his side, ready to
do his bidding, to assume any shape, and to serve him
in small matters and in greatest. In like manner man
has taken them all, from the precious gold to the worthless lead, and made them his servants. But there is one
among them, more highly gifted than all the others, of
fairest form, of strangest shape, and of rarest usefulness.

This is the metal which takes its name from the
winged messenger of the gods, and is known among us
as mercury, though its bright face and wondrous quick-
ness make it perhaps more generally familiar as quicksilver. These features were so striking and so exclusively peculiar to the ever-changing metal, that already
the ancients bestowed upon it like admiring names. To

the Greeks it was liquid silver; to the Romans, with a
poetry rare among the stern, matter-of-fact people even,
living silver; the latter name, however, seems in classic
writers to have been confined to the pure mercury found
in its brilliant beauty, whilst the former was reserved for
the metal when artificially produced. For the ancients
were already fully acquainted with the "Changeful
Damsel" among the stern metals, as an old alchemist
quaintly called it on account of its slippery, coquettish
nature, now alluring by its 'lovely beauty, now deriding
by its swift escape. Among its many forms, which it
assumes, is one called cinnabar, of a resplendent red,
which was well known more than four hundred years
before Christ, and found abundantly in Spain, where, by
one of those strange combinations produced by the universal rule of the Roman, Athenian philosophers acted as
officials in imperial mines. That able- but disorderly
writer, Vitruvius, confounds this cinnabar—on account
of its red color, in all probability—with the more familiar minium, an entirely different product, used to mark
certain passages in manuscripts and almanacs, and thus
become the ancestor of our miniature. He states, however, correctly the picturesque manner in which it presented itself to the eye of the astonished miner; for he
says, "When they dig minium, and iron tools wound
the rock, big drops of living silver flow from the place."
Pliny, in his more prosy manner, simply states that there
—in the mines of Spain—there is "a rock which continu-

ally sweats mercury, and which the Greeks call cinna-bar;" so that there can be no doubt as to the identity of this curious metal in Roman mines and our quicksilver. The question has, however, been raised more than once, because of its being so frequently mistaken for minium, and even called by that name. Pliny himself designates it thus wrongly in his interesting description of the locality, from which, in his day, the most valuable cinnabar was sent to the capital. "Rome," he says, "obtains its minium almost exclusively from Spain. The most famous comes from the region of Sisapo in Boetica. The mines belong to the Roman republic, and no other property is so jealously guarded as this. The cinnabar is not allowed to be prepared on the spot; but it is stamped as brute ore and sent to Rome, about ten thousand pounds a-year. In Rome it is washed and prepared, and a special law fixes the maximum price at which it may be sold by the merchants." Now Pliny's Boetica is the Andalusia of our day, with a part of Granada; and in this same district are still the famous mines of Almaden, the one great support of the Spanish crown, without which Spain would have long since been utterly bankrupt.

The subject was one of no slight importance to the Romans, for cinnabar was used largely for the purpose of painting. Its bright red adorned the statues of the gods as they were carried in solemn procession through the wards of the capital; and even the great generals, who entered the city in all the pomp and circumstance of a

full triumph, did not disdain its use. Sir Humphrey Davy
recognized its use even in some paintings of disentombed
Pompeii, and probably it served still higher purposes in
the mysteries of unholy worship. Mercury, as such, was
also well known already as useful for purposes of gilding, although the modern art of using it in the shape of
an amalgam was not familiar to the ancients. They con-
tented themselves with putting the mercury in a layer
on vessels and ornaments of silver and copper, and then
pressing thick plates of gold upon it, cementing the whole
together. Whilst our gilding, therefore, wears off even
by mere daily use, and when not exposed to the baneful
effects of wind and weather, we are told by the great art-
critic, Winkelmann, that antique gildings look now as
fresh and as beautiful as if they had just come from the
hands of the gilder. Hence they had also learned already to burn their magnificent dresses, embroidered
heavily with pure gold, and, by the aid of quicksilver, to
rescue the gold from the ashes.

How far it was used, even then, for medicinal purposes,
we can hardly decide ; for while some authors mention it
as an element in certain salves which were oddly enough
employed at festive meals, Pliny represents it correctly
as a poison, and objects to its use in medicine, even for
external application, as fraught with too much danger.
The confusion in the mind of these authors, when they
come to speak of the mysterious metal, is often amusing :
Pliny thinks it so poisonous that no vessel can hold it,

aware as he was, probably, that it cannot be kept in metal vessels, because it would at once form an amalgam with the metal; while, on the other hand, Dioscorides states that it was generally stored up in glass vessels, but that he has seen it also in boxes of lead, tin, and silver, which is simply impossible. Its fluidity, however, seems to have puzzled the ancients sorely; and the amazement of Vitruvius is comic in the extreme, when he describes how a stone, weighing a hundred pounds, put on a vessel filled with living silver, floated on it, without making an impression on the surface! This, also, is of course erroneous, for the stone does make an indentation, more or less deep according to its specific gravity, as mercury is only about thirteen times heavier than water; but it is curious that the same experiment, which so astonished the learned Roman, is in our day repeated daily for the visitors of the mines of Idria, whére huge stones are placed in the enormous iron kettles filled with newly-obtained mercury. It is strange that we find no trace in ancient writers of the preparation of artificial cinnabar, highly valued as this costly material was by the men of those days; but there is only one allusion found to what is called making mercury solid by the aid of sulphur, and that occurs unfortunately in the pages of the false Democritus.

This secret, like many others connected with our strange metal, was known only to the great race who kept all the valuable knowledge of the world in the days

of universal war, and through the well-named Dark Ages
—the Arabs, who also were well acquainted with the
deadliest form that mercury ever assumes, the so-called
corrosive sublimate, and described it as a violent and
acrid poison.

The alchemists, those noble searchers after truth, who
paved with their errors and bitter disappointments the
roads on which Modern Chemistry steps safely to the
great goal of Truth, surrounded mercury with a poetic
crown of glory. In their labors to wring from Nature
the secret of the philosopher's stone, and of changing
all viler metals into noble gold; in their efforts to realize
the existence of an elixir of life; in all their mad pursuits,
which blend the sublime with the ridiculous as no other
work of man has ever done, and contain, amid much that
is absurd, numerous traits of touching self-denial and
unsurpassed perseverance—in all of these, mercury was
the one great master among metals, without whose aid
nothing could be obtained. How these poor, ignorant,
but earnest and devoted workers worshipped the mysterious metal, and tried, by all the means known to their budding science, to force it into their service! Its change-
ableness was their despair. Not in vain had they named
it mercury, when they expressed the sympathetic relation
which they fancied to exist between the seven known
metals and the seven planets; whilst gold remained to
them the image of the bright sun, and silver the repre-
sentative of the pale moon, quicksilver bore the likeness
14

of the messenger of the gods with his winged foot and mobile mind. So they fasted and prayed, and chastised themselves into a proper frame of mind, to become masters of the volatile servant; and then with exalted hopes and a glance to the Master of all things, they heated and cooled, digested and distilled, analyzed and amalgamated the unlucky metal, in order to find the animated mercury, as they called the future substance, which was to make the philosopher's stone, and the more mystic philosopher's mercury, from which they expected still greater but unknown wonders. They believed even in predestination as required for the happy issue of their work. Alas! they were predestined only to work out all the errors of human knowledge, and to clear the way for their successors in ages long to come. Their success was limited to chasing the metal from one shape to the other; now changing from living silver into the red lion, then into cinnabar, the dragon's blood, and back to the milk of the black cow. Even Geber, the acknowledged master of many a science, became, as the unfortunate author of the first book on chemistry ever written, a byword among men. Dr. Johnson tell us how, on account of his uncouth language in this work, his name has been transmuted into Gibberish for the use of indignant English tongues. To him mercury was the dearest among the rare and aristocratic substances with which he loved to deal, and, with sulphur and arsenic, one of his three elemental chemicals, of which all metals on earth were made. He dwelt with intense pleasure upon the fact that even gold, the sovereign of them all, with its superior weight, its passing beauty, and its triumph over fire, was dissolved by mercury, and swallowed up by its bright globules as easily as sugar in water.

The alchemists failed in their end, but they have taught us much about quicksilver. For we would err sadly, if we were to look upon them as lost in error altogether. If Wisdom in their days wore the fool's cap, there were wise things said and done even in her wildest vagaries; her secretary, as he has been called, Common Sense, made notes of the good, and all was put down in a kind of short-hand, strange and odd to our ear, but intelligible to the initiated. The vocabulary was made awful and hideous on purpose, to keep off the profane; but fair Science came out at last unscathed, for Truth cannot be destroyed nor concealed; and thus it appeared, that philosophy, like the toad, ugly and venomous at first sight, bore "a precious jewel in its head." It was not for such triumphs, it is true, that the great Dominican, Thomas Aquinas, was enrolled among the saints of the Church; but it is a fair question among men of science, whether this canonized alchemist achieved any greater triumph on earth than his discovery of the word amalgam, and its meaning. Now we all know that mercury bears such marvellously strong affinity to certain metals like tin, lead, and silver, that it opens them up, so to say, and forms with them a homogeneous liquid or paste; but Aquinas was the first to ascertain and state this fact in a manner which made it useful to science, and to call the compound, that resulted from the mixture, by the name which it still bears.

Quicksilver is not a common metal, but found only in a few portions of our globe, and then in various forms. Here, it surprises the miner by suddenly leaping forth in bright, silvery globules, and running fast to hide in dark corners, now scattering into almost invisible atoms, now running together and forming large, lustrous balls. Then, it appears as cinnabar in fibrous or ill-shapen masses, sometimes crystallized and sometimes hid under a dark, unseemly covering, but occasionally shining forth in brilliant red, with the splendor of the diamond. Then again it conceals itself carefully under the form of some other metal, as native amalgam, or, in rare cases, is found as ready-made medicine in the form of calomel.

China and Japan produce some quicksilver, for they send to the markets of the world the best cinnabar which there appears; but to this day little is known about these mines and some others in Thibet. The so-called virgin cinnabar, which is mainly imported from Asia, is made from the rare specimens of native cinnabar found there and ground fine; it is by far the most superb in color, and the most highly prized by artists and manufacturers; the larger portion is probably made artificially in China as in Europe, and for that purpose large quantities of mercury are actually carried back to the Celestial Empire.

Mercury is found in the Venetian Alps, where new mines have quite recently been opened with a promise of great gain; in Bavaria and Bohemia, and a few other localities, which have, however, been abandoned of late on account of their small yield, with the exception of here and there a small mine. There is a curious theory about this decline of formerly productive mines entertained by older authors. They state that mobile quicksilver cannot rest, even when rockbound, but ceaselessly works its way upward, and from great depths penetrates, by the process of sublimation, into the veins and crevices of overlying formations. This theory is supported by the fact, that in almost all the formerly rich mines, mercury was actually found close to the surface, often immedi-. ately under the turf; whilst no ore has of late been discovered at a lower depth.

On the whole, there are only four important points on our globe where mercury is mined to advantage :· in Peru, in Austria, in Spain, and in California. Of these, the smallest production is that of Huancavelica, in Peru, where cinnabar appears impregnated in layers of sandstone and limestone. It is curious how the vanity of man here, as in Spain, first led to the discovery of the precious metal. .For as the great Triumphator Camillus painted himself red with minium, so the Indians of Peru used to adorn themselves on festive occasions with the same color, obtained from the same metal. The first regular mining dates only back as far as the year 1566, when the

Spanish Government caused the province to be examined, and mercury was found in numerous places. When Peru became independent, the mine of Huancavelica became, of course, national property, and was farmed out by the Government. It is the highest mine on earth, being fifteen hundred feet above the Peak of Teneriffe; but it produced a rich harvest, until the madness of an official nearly destroyed it forever. The development of the silver mines of the republic, and the wasteful manner of using mercury in them, had led to an increased demand for the metal. This induced an unlucky superintendent, goaded on by his superiors, and anxious to distinguish himself, perhaps also to add to his riches, to order the massive pillars of valuable ore to be pulled down, which had been left standing in order to support the enormous weight of the upper surface. The haste to be rich had its unvarying effect: the rock came down, the mines were destroyed, and owner and agent alike were severely punished for their insane cupidity.

Far away in the heart of Europe and the centre of the Julian Alps, there lies a valley of wondrous beauty; huge walls of bare rock rise to vast height all around, and shut off the secluded plain below from the rest of the world. The upper parts of the mountains are again crowned with grand old pine-forests; below the rocks, spreads a beautiful carpet of green meadows and magnificent woods, while here and there tall masses of rock jut out into the valley, looking defiantly down upon the peaceful scene below, and crowned with chapel and church. A poor peasant—so goes the legend—once came here to select some timber for the wooden-ware he was making, and placed a few tubs into a well to be seasoned over night. What was his surprise, when he poured the water off next morning, to find at the bottom a glittering mass of silvery metal! Soon skilful miners and cunning goldsmiths came from Italy, and the place became known; but only to be made the scene of strife and bloodshed. Now the Venetians would fall with an armed force upon the German miners, and now the great Maximilian would send troops to drive out and destroy the covetous invaders. For centuries the mines have produced most abundantly, but of late they have become less profitable, and the Austrian Government, always wanting money, is anxious to sell them. The ore is here roasted in extensive works, and the smoke, which contains the volatile metal, is carried into enormous iron retorts. The cast-iron pipes are forty feet long and three feet wide, suspended in the air, and resting only upon a few slight supports. As the constant smoke keeps them too warm for condensation, a little aqueduct, following above, trickles continually cooling showers upon them; once a year only the fires are allowed to go out, to clean and repair the whole apparatus. The soot in the cylinders, a loose, black, fatty substance, contains the mercury in countless tiny globules, some of which run out in beautiful silvery whiteness; others have to be whipped from their un-

sightly retreat; the soot is beaten with small brooms, and
soon the silver snakes are seen to glide out of the dark
mass, as if anxious to escape in all directions. Then the
glittering mass is gathered in sheepskins, tanned with
alum, or in cast-iron bottles of enormous size, to be sent
all over the world.

Sad, however, is the penalty which the vicious metal
exacts from those who thus force it to the light of day.
Quicksilver evaporates at a temperature lower than that
which we maintain in our houses, and its vapors are
poisonous. Hence all miners pay with their health; they
become feeble and nervous, their whole system becomes
deranged, perpetual trembling seizes their limbs, and
they die at an early age. When murcury escapes by
chance, it murders like an assassin in the dark. Thus it
happened in a vessel which in 1820 came to the Spanish
coast with some mercury in its hold. By an accident
the quicksilver ran out of a few rotten bags, and found
its way into the hold; soon every piece of metal in the
ship was covered with a mercurial coating, and every
man on board was salivated violently, and sick unto
death. The same tragedy occurred in Idria on a larger
scale. In the year 1803 foul air set the mines on fire, and
the mercurial vapors developed on that occasion poisoned
thirteen hundred workmen, the larger part of whom never
recovered. The water, by which the fire had been
quenched, was pumped into the river Idria, and was still
so laden with deadly fumes that all the fish were killed,

except the eels, who, being proverbially used to being roasted alive, defied even the poisonous metal.

On a similar occasion, in Spain, a pious Franciscan monk seized a crucifix, and bidding the frightened workmen to follow him into the burning mine, went in to put out the fire : neither he nor any of his devoted men were ever seen again. This occurred in the famous mines of Almaden, which lie amid the Black Mountains of La Mancha, so dear to all lovers of Cervantes as the scene of the inimitable exploits of Don Quixote. It has its name—the mine—from its first masters, the Arabs, who, in the course of time, fell heirs to the Roman State property, and worked it with brilliant success. One of their Caliphs had in his gardens at Cordova a gigantic shell of porphyry, filled with glittering quicksilver, which was evermore flowing out and in. It stood in a pavilion, the sides of which were lined with ebony and ivory of such exquisite polish that, when the rays óf the sun fell upon them, their splendor dazzled and blinded the eye. But when great guests were to be shown the marvels of the palace, an Arabic manuscript says, the Caliph ordered all the doors around to be opened; the full sunlight then shone upon the ever-flowing mass of silver, and the reflection fell on the eye of the beholder like living flashes of lightning, and the pavilion seemed to toss like a vessel on the stormy waves.

The indolent Spaniards have rarely attempted to work their magnificent mines themselves, but farmed them

14*

out to bankers and companies, mostly of foreign race. Among these was the great house of Fugger, those grandest of merchant-princes in the small town of Augsburg, one of whom could haughtily warm the Emperor Charles V. by a fire made of his own obligations and evidences of debt to the great banker! They held the mines for more than a hundred years, and brought large numbers of German workmen there, whose industry and skill soon raised the production immensely. In 1835 they were pawned, in like manner, to the Rothschilds; but at present they are worked with great energy and by means of the best scientific aid, at the expense of the Government, which derives by far the largest portion of its income from this locality. The town itself, with its clean, straight streets, and tidy, well-kept houses, presents a very different aspect from the other miserable villages of La Mancha, and the surrounding country; though sterile and stern like that of most mining districts, is not devoid of beautiful mountain scenery and extended views. Even the entrance to the mines differs altogether from that of similar works elsewhere. From the level valley a long tunnel-like shaft leads to the very heart of the mine; it is built of massive hewn rock, wide enough for carts with two horses abreast, and has granite sidewalks; at the end you come to steps and stairs, which lead to the lower parts, where you find yourself immediately beneath the town of Almaden. The working was formerly done by criminals condemned to hard labor for life. From their prison, which still stands there, a relic of former barbarism, they were led in the morning by a subterranean passage to the mines, and back again at night. Thus they literally never saw the light of day; after a few years their health failed, the poisonous vapors nestled in their system, and they died, hailing the King of Terrors as a welcome friend. This cruelty drove them at last to despair: in the beginning of the last century they set the wood-work, which then lined the mines throughout, on fire, and thus made them inaccessible for years. Now, none but skilled laborers are employed, who work only six hours a day, and are well paid. Yet they also soon succumb, for the air is so deadly that no animals live down there, not even spiders; and the rats, who alone were able to resist the vapors, have disappeared since the last fire.

The scene below is beautiful. Where the work is going on, vast masses of cinnabar, dark-red and sparkling with unearthly splendor, hang on the walls; here and there crystals of marvellous beauty shine from between the dark rocks, and in many places cavities and crevices are filled with the pure metal; so that, under the miner's tool, as the rock breaks asunder, silvery drops as large as a pigeon's egg suddenly roll forth, and, leaping on the ground, fall into a shower of resplendent beauty. The ore is subsequently distilled by means of enormous fires, for which prodigal Nature furnishes bountifully the material; for all the mountains around,

for miles and miles, are covered with a cistus, an ever-
green shrub, which, at the proper season, covers the
country with a carpet of white, and whose resinous
branches burn with a heat equal to that of the best coal.

The richest of all regions, however, is the youngest—
California. Here, where Nature seems to have scattered
with unlimited liberality her greatest treasures broadcast
over the land, vast stores of mercury are found, the most
important, if not the most valuable of the mineral pro-
ducts of this wonderful country. To the four mines,
which already produced more quicksilver than all other
countries together, there have quite recently been added
two new ones, in Chapman Valley and Pope Valley,
which give promise of a vast increase of the wealth of
California, and have completely changed the commerce
of the world. Formerly, mercury was exported from
Europe; now, America is not only independent, but actually sends it back to the Old World, and enables men
there, as well as in Peru, to work even the poorest of
silver ores. Thus silver, which had remained behind
gold, since the great discoveries in California, Australia,
and Asiatic Russia, can now be obtained again in larger
quantities, and, thanks to the new supplies from the Pacific, the balance is likely to be restored.

If we ask, finally, what use man makes of the wonderful metal, so beautiful in appearance, so deadly in its
effects, and so highly valued for its services, we find it
nowhere employed for itself, like the more precious met-

als, but an invaluable aid to man in various ways. Its
very dangers are converted into healing powers, and, as
calomel, it is used with surpassing effect, either directly,
mixed as in bitter irony with tender roseleaves, or in
numerous combinations with other substances. As corrosive sublimate, it enters into countless salves of more
doubtful usefulness; and as mercury itself, it is infused
into anatomical preparations, to preserve them for purposes of study. No chemical laboratory can, of course,
dispense with its valuable services, were it only in the
form of a bath, to catch gases. The manufacturer
employs it largely for dyeing and similar purposes, and
the man of science learns to value it as a friend above
all others in the thermometer and barometer. He
wanted to measure that imponderable and yet universally present substance, heat; and quicksilver willingly
helped him, because of its exquisite susceptibility, and
the readiness of expansion which it alone could show, as
the only fluid-metal on earth. Man wanted to weigh the
very air he breathes, and quicksilver again offered to do
it, as no other fluid is so heavy, and a column of twenty-
eight inches of mercury suffices to show what water
could have done only in a tube of thirty-two feet. By
the aid of these instruments, man can now measure the
warmth of his room as well as that of whole zones; he
can adapt the temperature of his bath to his wants, and
of vast buildings to purposes of brewing and distilling.
The mercury aids him in measuring the height of moun-

tains, and warns him of approaching storms; it counsels him in his work in the fields, and on his voyages over the oceans; it aids the engineer in his levelling, and the philosopher in his subtlest researches.

In other cases it is mixed with sulphur to make artificial cinnabar, and immense quantities of mercury are annually consumed for this purpose. It is ground to extreme fineness, and then comes into the market as vermillion, which is highly valued in the arts as a pigment for the purity and permanency of its tint. But by far the largest proportion of all the mercury found on our globe is sent by man in search of the precious metals, which force cannot bring forth from their hidden recesses, and which now, with the aid of the alluring powers of quicksilver, are tempted to the light by its gentle persuasive power. For it possesses a truly wondrous power to compel gold and silver to leave their natural form, and to combine with itself; and hence the poorest ores, which by no process of beating and heating can be made to surrender the treasures they hold, are covered with mercury, and behold, the insidious friend winds its way into every chink and crevice, and licks up the precious metal wherever it can be found. Then the whole mass is exposed to a fierce heat, the volatile mercury is forced to surrender its prey; it rises in vapor and smoke, and leaves the glittering gold and silver in almost perfect purity behind. The same process of amalgamation leads to gilding, for here also a mixture of gold and mercury is put on silver, copper, or brass, and the metal is then heated; the mercury again escapes, and the gold remains firmly fastened to the inferior metal, which is said to be " fire-gilt." The process is obnoxious on account of the injurious vapors which it devolves, and largely superseded by the galvanic method, which is safer and cheaper; but the gilding wears off sooner. Not less beautiful is the method by which mercury serves to make mirrors. Tin-foil is spread on a hot slate-plate, the quicksilver poured on it, and then the plate of glass pressed carefully under the surface of the mercury, so as to avoid the particles of dust which always settle there. The glass is then heavily weighted, the quicksilver runs off in glittering rills, and the mirror is perfect. But here also the danger has driven man's ingenuity to rid himself of the beautiful vassal, with the fair face and the fatal poison behind it, and nowadays mirrors are simply silvered.

XII

THE EARTH IN TROUBLE.

" Thou sure and finest earth."—SHAKESPEARE.

THERE is no mistake about it ; our mother Earth is in serious trouble, and her wisest children are at a loss how to account for her sudden restlessness. There are all the signs of feverish excitement—great heat, strange utterances, and violent convulsions. A summer so hot as to become unusually destructive to human life has been followed by an unnaturally mild winter all through the temperate zone of the globe, and even the instincts of the brute creation seem to have been at fault for once. Terrific upheavings have terrified man, now breaking forth through the craters of active volcanoes, and now raising huge portions of firm land by fearful earthquakes. Is it a wonder that when our mother Earth is so evidently in trouble, her children also should be sorely troubled, and thoughtful minds should look once more for the speedy coming of the end of all things ? When the Apostles themselves expected to witness the coming of the Lord, and a Luther could firmly believe in the near approach of the Last Judgment, we may well bear patiently with credulous Millerites, sitting in their white robes high up on broad branching trees to ascend the more promptly to heaven, and with all the sorrowful minds who in our day yearn, with the whole groaning creation, for speedy redemption!

Nor can we withhold our sympathy from those who describe, with feelings of indelible awe, what they suffered at the time of their first experience of an earthquake. While a bright sky and brilliant sunshine are flooding the exuberant beauty of a tropical landscape with gorgeous lights, and all Nature seems to enjoy in perfect peace the luxury of happy existence, they suddenly feel rather than hear a low, rumbling noise, which seems to rise from the very lowest depths of the earth. And all living beings, men as well as animals, are of a sudden filled with a strange anticipation of evil coming, vague, but sickening, and unconquerable by any effort of will. Before the mind can well judge of the strange and unwonted sensation, there comes long, subterranean thundering, clap upon clap, rolling nearer and nearer, and at each successive shock the heavens and the earth seem alike to shudder at the fearful approach of an unknown power. Everything is shaken to the foundation; glasses and crockery-ware sound as if frolicking spirits were playing with them; bells are set ringing by invisible hands, doors open by themselves, and no one enters, the houses begin to groan and to crack in all their joints, and lean, like drunken men, first to one, and then to the other side. Tall steeples sway giddily to and fro, and lofty arches in cathedrals and churches press out the keystone and come crashing down, burying thousands of terrified men, who had come to the sacred building to invoke help from on high, when all upon earth had left them helpless. All who can escape rush forth from beneath treacherous roofs, but out there it feels as if even "heaven's vault should crack;" the danger is not over, for the very soil beneath their feet swells and sinks like the waves of the sea, huge chasms open here and there, and dark abysses swallow old and young, rich and poor, without distinction and without mercy.

At last the soil begins to subside into the wonted quiet, and at the same moment, a tall, conical mountain—sometimes in sight of the panic-struck multitude, and sometimes at a distance of hundreds of miles—opens a wide, gaping mouth near the summit, and a power, which human ingenuity has as yet found no standard to measure, sends forth a magnificent bunch of bright flames, mingled in strange anomaly with streaming vapors, rocks ground to atoms till they resemble ashes, and vast masses of a glowing substance, which are flung, jet after jet, till they seem to reach the welkin. And, high up in the air, the fiery bouquet, grandly beautiful in spite of its terrific nature, spreads out into an immense canopy, an ocean of clouds dark above, but shining in incomparable splendor

below, where the fire from the crater illumines it in richest glory, while flashes of lightning play unceasingly to
and fro, and the half molten rocks rain down upon the
earth, bursting and breaking like masses of brittle dusky
glass. At the same time a torrent of ashes falls like a
burning rain of withering fire upon the wretched landscape, and in an instant the whole region, for miles and
miles all around, is covered with a weird shroud and
sinks for ages into death-like stillness !

But troubled Nature has not exhausted the efforts yet,
by which she seeks relief from the mysterious suffering
which she seems to undergo in the dark recesses of the
earth. From the crater's brink, or from a sudden opening in the sides of the mountain, there comes gushing
forth a broad stream of fiery lava, and hurries, as in
furious madness, down the steep sides, carrying the
torch of destruction to the forests, which in a moment
flare up in a bright blaze, to fertile fields, changing
them instantly from lovely pictures of peace and promise
into desolate deserts, to lofty walls and solid mansions,
which crumble and fall at the magic touch, never to
rise again, and finally to the silent sea, into which it rolls
its fiery waves with a fearful hissing and screeching,
bringing even here death and destruction to all that lives
and moves in the life-teeming waters.

And, as if the measure of horrors was not full yet, and
overburdened Nature must give vent in new forms to
its unbearable burden, the heavens darken, till night

covers the earth, and a deluge of waters descend in vast
sheets, flooding the fields that had barely begun to
breathe once more freely, and mingling in horrible friendship with the masses of black ashes, so that the dark
hideous slime rolls in slow, but irresistible waves, over
town and village, and fills cellars and rooms and streets
and the very temples of the gods with its death-bringing
horror. And not unfrequently the sea rushes up to meet
in fatal embrace the waters from the clouds; trembling
under the weird excitement and coming up in fierce,
spasmodic jerks as the convulsions of the volcano near-
by shake it with sympathetic violence, it breaks down
the ancient landmarks that have held it in bounds for
countless ages, and retreating after a while with overwhelming violence, it bears the few survivors from the
fury of fire into the fatal abyss of the ocean.

Amid such horrors the bravest of men loses heart, and
with all his heaven-appointed powers he feels like a
helpless infant. The brutes of the forest, the lion and the
panther, forget their nature, and come from their dark
dens to join in strange, new-born friendship, the flocks
of peaceful cattle, and to seek with them, driven by an
irresistible instinct, the shelter of human habitations
and the protection of man. Eagles and vultures come
down from their unseen paths in the clouds and their
lofty eyries, and sit, marvelling and trembling, by the
side of pigeons and common fowls in paradisaical peace.

It is this unique and uncontrollable sensation, felt

when the material world makes for a moment its full
dominion known, and claims our earth-born nature as its
own, which has, no doubt, led, from of old, to the almost
unvarying creed of men, that the world will come to an
end by fire. The Chaldæans, it is true, coupled the
power of water with that of the burning element, and
believed that the world would be destroyed by fire, when
all the stars should meet in the constellation of Cancer,
and once more by water when they meet in the constellation of Capricorn. The Parsees, worshippers of fire,
have a similar doctrine, according to which the world will
last twelve thousand years, after which Ahriman, the Evil
Spirit, will set it on fire by means of a comet, and, after
a thorough purification, re-create it with Ormuzd, the
Spirit of Good. Even the Orphic poems, of which nothing is left beyond a few quotations and allusions, are said
to have sung of the end of winter in a great deluge, and
of the end of the world's summer in a great conflagra-
tion. It is well known, on the other hand, that the Mosaic Genesis, based, perhaps, largely upon the impression
produced by the annual inundations of the Nile, admits
of only one creative principle, that of water, which
" brought forth " all things living but man, and hence
laid the foundation of that system which is still warmly
defended by the Neptunists of our day.

It is interesting to observe how, in a similar manner,
the Greek Heraclitus drew his views from his observation
of volcanic symptoms, and based upon them his theory,

that the world not only owed its origin to fire, but was
to be periodically purified and renewed by vast conflagra-
tions.   Fire was, to him, the only unchanging and
everlasting element, and to its benign influences he was
disposed to ascribe all that befalls our globe under the
direction of relentless Fate.

Nor can it be doubted that the same impression led
originally to the almost sublime conceptions of the lower
regions, which we find in Hellenic legends.   It is well
known that they placed their Tartarus far down in the
bowels of the earth, and represented it as an enormous
abyss filled with eternal fire.   The very position of the
entrance to this lower world, in Southern Italy, points to
that connection, as the active volcanoes of that region
had, no doubt, originally suggested the whole concep-
tion.   Far down, below those favored plains, they imagined the realm of Pluto, and looked upon Mount
Vesuvius and Mount Etna as the colossal chimneys,
giving vent to the smoke of the fire at which the Cyclops
were forever busy forging the lightnings of Jupiter.
How deeply rooted these fantastic and yet beautiful no-
tions were in the minds of nations, we may judge from
the fact that two hundred years after the rise of our
faith the Roman historian, Dion Cassius, could still
soberly speak of enormous giants rising from Mount
Vesuvius, and scattering, amid the appalling sound of
infernal trumpets, ashes and rocks over the blooming
fields of Campania and the fair cities of Pompeii and

Herculaneum! Our own Christian faith, finally, teaches us of the final destruction of our globe by the same terrible agent, when "the heavens shall pass away with a great noise, and the elements shall melt with fervent heat ; the earth also, and all the works therein, shall be burned up."

If it is strange to see how universal this fire worship is, which ascribes to this element, above all others, the power to create and to destroy our world, it is not less striking to observe to what eccentric views the same conviction has led both ancient and modern inquirers. Thus Aristotle even was fond of imagining that the earth might be a living being, which changed like man on the surface, only at much longer intervals. He knew perfectly well that certain portions of land would gradually be covered with water, while parts of the sea would be laid bare and change into fertile lands ; he knew equally well the origin of volcanic islands, and describes correctly the sudden rise of Hiera, in the Pontus, which was born amid a fearful upheaving of the earth, its bursting open in the shape of a great crater, and the subsequent lifting up of a high mountain. All these phenomena were, to him, evidence of the inner life of the earth, which, he thought, manifested itself mainly by fire. Strabo went even beyond him, and while ascribing, with his illustrious predecessor, all earthquakes to the efforts made by masses of heated air within to break through the crust of the earth, he discerned the correct origin of the great

changes on the surface, and, for instance, saw in Sicily
only a portion of the main land, which had been detached
from it by a violent volcanic upheaving.

The Romans did nothing for the better knowledge of
Nature; their thoughts were exclusively given to the
Empire, and social problems monopolized their attention.
For centuries, therefore, natural science made no progress,
and earthquakes were readily ascribed to rebellions in
the demon world below, and volcanic eruptions to the
impatience of chained spirits. Then came the rule of
Neptune, when Vulcan was dethroned for a time, and
all the great symptoms of life, which our mother Earth
gives forth from time to time, were explained by the
agency of water. Descartes was the first philosopher
bold enough to leave the beaten track, and to plead once
more the cause of fire; he openly declared his conviction,
that the earth had once been a fiery meteor, like so many
others, fragments of that original solid matter which had
been set in furious motion by an Almighty hand, and
when heated by the terrible velocity with which it re-
volved in infinite space, divided into suns and stars.
His doctrine was, that the crust of the earth had afterwards gradually cooled off, but that in the interior there
was still a vast central fire, which every now and then
spontaneously bursts forth in eruptions and earthquakes.

The greater philosophers of later ages followed in the
wake of Descartes, without adding strength to his argu-
ments or facts to support his theory; it was only when

the three great naturalists, Pallas, Saussure, and Werner, collected a number of carefully-made observations, that speculation was changed into conviction, and brilliant suggestions were tried and proved in the alembic of stern logic.

They did not change the original theory, but established it on a solid basis. They found, what we still believe, that the sea of liquid fire beneath the thin crust of the earth, on which we dwell with fancied security, is in a state of perpetual excitements, and hence continually presses or rises against the surface. When it touches it, we have an earthquake; when it is sufficiently excited to break through the crust, it forms a volcano.

According to the most recent theories, however, another new element has been added to these explanations of the inner life of our earth. We have learnt that it is not fire simply, which produces the agitations, but the same power which raises the waters of the ocean at regular intervals. It is well known that the tides are the effect of the attraction exercised by sun and moon, and that they are highest in the form of spring tides, when sun and moon combine to attract the waters. It is believed, now, that the liquid matter in the interior of the earth obeys the same laws of attraction, and rises and falls with the outer liquid, as the crust of the earth is, relatively speaking, a mere thin covering, unable seriously to diminish, much less to check, the powers of attraction exercised by the two great heavenly bodies. This view is confirmed by the fact that earthquakes are most frequent when the tides are highest. Volcanic eruptions are, of course, effects of the same commotion below; they only save vast regions of land and water from being thus convulsed, by offering an open vent to the gases developed below.

All this newly acquired knowledge, however, does not yet help us to avert the fearful destruction which generally follows the outbreaks of the hidden power within the earth. In vain do we see vast plains laid waste forever by the death-bringing substances ejected from hideous mud-volcanoes; in vain do towering mountains rise where formerly the eye swept over level lands as far as it could reach; in vain, even, do we descend to towns which once overflowed with life and exulted in their splendor, and which now are sad and silent, buried for ages and ages below the surface of the earth. At each new return of the terrible calamities attendant on such convulsions, we stand anew aghast, and feel how utterly helpless we are, how utterly ignorant even of Him who "laid the foundations of the earth," and who alone knows "whereupon are the foundations thereof fastened, or who laid the corner-stone thereof, when the morning stars sang together, and all the sons of God shouted for joy."

So we have felt again during the present year. For our mother Earth has been in great trouble during the last twelve months, and perhaps it may not be amiss briefly to record here the symptoms which make us aware of the terrible commotion which has apparently destroyed the peace ordinarily reigning within our globe.

Earthquakes have taken place in the West Indies and in South America, such as belong to the most terrific catastrophes recorded in the annals of the earth's history. Since the day on which Lisbon was swallowed up with thousands of helpless victims, and the calamity at Lima in 1746, since the South American coasts were devastated in 1797, and Caracas was utterly destroyed in 1812, no such overwhelming misfortune has befallen that doomed locality.

The air, we are told, had been for several days so hot and oppressive that experienced natives foretold a volcanic eruption. On the 16th of August (1868) news was received in Valparaiso, that in several ports of Chili the sea had risen and overwhelmed the coast for fifteen miles, so as to wash away houses and magazines, and to land vessels high and dry far inland. Three days before the earth had begun to heave, and regular earthquakes had taken place at Callao, returning at intervals of five minutes. Enormous crevices opened, houses fell, churches crumbled to pieces, and men and brutes alike were frightened by the unseen enemy. The whole West coast, as far as high up in Peru, was thus shaken, and at various places the sea had made inroads upon the firm land. At the very first shock a number of towns in the interior were levelled with the ground, and ancient cathedrals, that had stood like unchangeable landmarks for hundreds of years, were changed into heaps of ruins and rubbish. More than thirty thousand human beings perished in a day, and the loss in material and merchandise is beyond all calculation. An enormous spring tide followed the earthquake, and overwhelming the frail bulwarks of a low coast, flooded the land far into the interior. Large vessels were thrown from their anchorage, and landed far up the country. A second gigantic wave, stretching a hundred miles north and south, rose from the ocean, and fell with crushing power upon the ill-fated coast. Three war steamers were thus destroyed at Arica alone; among these our own ships, the "Wateree" and the "Fredonia," the latter with nearly every soul on board. An English steamer, the "Santiago," escaped by a marvel. She was apparently secure, riding on two powerful anchors; but suddenly a concussion was felt, which made the large ship twist and turn as if she were made of India rubber. All the passengers were tossed up to the height of two feet, and then fell flat down; at the same moment the heavy cables snapped as if they were thin wires, and the vessel was swept by a receding wave into the sea. Fortunately, they had steam up, and tried to gain the offing; but the next moment a second wave came, and drove her irresistibly towards the rocky shore. All faces were deadly pale, and the captain gave up all hope. But oh, wonder! the wave lifted the ship high up, and safely carried her on her gigantic shoulders across the rocky barrier, letting her gently down into an adjoining bay, from which she could subsequently escape into the open sea! Where the town of Chala stood, the ocean now floats heavy vessels, and Iquique was destroyed first by an earthquake, which lasted uninterruptedly for four minutes, and then by a wave of sixty feet height, which suddenly approached the land like a solid wall, and then fell, crushing all that it found in its way, together with more than a hundred men. Arica was so utterly destroyed, that even the places where certain prominent houses had stood could no more be found; and the unfettered fury of the waves had lifted up heavy guns, and borne them scornfully from an island battery, far out at sea, to distant inland hills. But the concussion itself extended far beyond the usual limits. Most powerful, as was natural, in the centre of the commotion, the volcanic region near Arequipas, where the famous group of snow-covered volcanoes form so striking a feature of the landscape, it was felt for a distance of six hundred miles, both of latitude and of longitude. Electric lights were seen in the air at different places—an entirely new phenomenon, not hitherto observed in connection with such events—and even the famous Tambo d'Apo, a house of refuge on the very summit of the Cordilleras, was so violently shaken as to crumble into dust.

It appears, however, upon a careful sifting of the evidence, that, after all, the earthquake itself did less harm than the sea. The enormous waves, which disobeyed the command, "Hitherto shalt thou come, but no further, and here shall thy proud waves be stayed!" carried utter destruction wherever they touched man, or the work of man. Houses and churches, fields and forests, all were literally swept away, islands disappeared, mountains were levelled, and dire desolation imprinted on the scene of abundant prosperity. But the worst was, as ever, the passion of man, unchained at a moment when the fury of the elements seemed likewise to be unfettered. Accident in some cases, fell purpose in many more, set fire to buildings, and soon large portions of the doomed towns were ravaged by fire and water alike! The excited populace fell with savage eagerness upon the stores of liquor exposed in cellars and warehouses, and soon hell itself seemed to be let loose. The scenes enacted in some of the unfortunate towns are beyond the powers of description; men beastly drunk lay by the side of those they had murdered, and the demoniac powers of the earth, set free by an unknown hand for a moment, seemed to have roused with fearful success the demoniac instincts in the heart of man.

These terrible occurrences were soon followed by similar calamities in the northern part of our Continent. An enormous spring tide, on the 15th of August, terrified the people on the Californian coast, rising to a height of over sixty feet, and washing away fields and gardens for miles. The earthquakes of Peru seemed gradually to have worked their way northward; for in the middle of October heavy commotions were felt, and on the twenty-first a violent earthquake shook San Francisco. The eastern part of the city was sorely tried ; many houses fell, others cracked from the foundation to the roof, and hardly one could be found that had not suffered some injury. As the shocks continued, all business was suspended, and a few cases of death soon caused universal consternation. Half of the population ran into the streets, but here also danger and death even lay in waiting ; for in several districts the earth opened, and jets of water leaped up to a height of several feet, while in other places the ground suddenly sank several inches. All the clocks stopped at the moment of the first shock, and the telegraphic wires were so much injured that no communication could be had for some time. The City Hall was a complete ruin ; the courts could not sit, and the prisoners were sent to the county jail ; the patients at the navy hospitals had to be removed, and the Mint was closed, until it could be fully repaired. Here, also, the shocks extended to a very unusual distance far inland ; and as they were felt at sea by sailors, who for a moment thought the vessel had touched a submarine rock, so they amazed miners in the interior, who expressed naïvely their indignation at such "indecent behavior of the old Earth."

The Pacific Ocean had a large share of the fearful commotion which caused such sad destruction on the adjoining Continent. Already in March a hundred earthquake shocks had been felt in the volcanic island of

Hawaii, connected with an unusually violent eruption of the far-famed Mauna Loa. Here also the earth opened in many places, and a tidal wave, sixty feet high, rose over the tops of lofty cocoa trees, and swept houses and gardens, cattle and human beings before it with irresistible violence. A terrible shock prostrated houses and churches, while the crater of the great volcano was vomiting fire, rocks, and lava, and a river of red-hot lava flowed for nearly six·miles to the sea, destroying everything before it, and forming a new island far out in the ocean. In April, still more violent shocks occurred, during which the swinging motion of the earth was so dreadful that no person could stand, and old and young were made deadly sick. At the same time tall hills were upheaved, and the tops detached, being thrown down into the valleys below, while out at sea new islands arose, several hundred feet high, and emitted for days a column of steam and smoke.

A few months before similar phenomena had been noticed in British East India. Earthquakes were felt, though only slightly, in various districts of the northern provinces, and what was most curious, they seemed to be strictly limited to a narrow line running northeastward. In one region, near Chindwana, an entirely new feature was superadded to the more familiar horrors of such catastrophes. Each shock was preceded by a heavy detonation, as if a whole park of artillery had been practicing in the neighborhood. Special agents were despatched to observe the phenomenon, which the natives had reported for several months already, and they heard the same noise, and felt immediately afterwards the usual vertigo produced by slighter earthquakes.

Europe has escaped these disasters, with the exception of such slight shocks as were felt, at intervals, in the United States also, but without producing any other impression than that of a very unusual state of commotion in the interior of the earth. Premonitory symptoms had shown themselves already in the preceding year (1866) in the Azores, when violent earthquakes shook the islands, and the sea rose, between Terceira and Graicoas, amid terrific detonations, and cast up jets of water to an enormous height. In June, stones began to be mingled with the vapors, and the amazed spectators beheld the ocean in commotion, throwing up enormous blocks of stone amid dense vapors, and emitting so strong a sulphurous odor that it could be borne only with great difficulty near the shore. It is probable that the old world was saved the fatal effects of violent earthquakes by the readiness of Mt. Vesuvius to give egress to the rebellious powers from below. The ancient volcano had, early in the year 1868, already given signs of increased activity, and whilst the flow of lava had ceased, the last-formed cone began to give out thick black clouds of smoke, in which brilliantly glowing masses of rock were occasionally seen. On the first day of October the marvellously ingenious instruments de-
15*

vised by Lamont, began to indicate a disturbance below
the soil, and a displacement of the surface, and the volcano became noisier than before. A small cone opened
next, at the side, from the summit to the base, and lava
issued forth, covering the former summit of Vesuvius.
It was here, for the first time, that the renowned director
of the Seis observatory, Palmieri, observed the periodic
nature of these volcanic eruptions. Each day the lava
would cease to flow at certain hours, and begin anew
after a short interval; twice a day, also, the active cone
would make an increased noise, and throw out its projectiles with greater violence. The correspondence thus
shown between the volcanic ebb and tide and that of the
sea was still further illustrated by other changes in the
flow of lava, by certain phenomena occurring at greater
intervals, which careful observation proved to take place
in unfailing sympathy with the motions of the moon.

In November, the mountain became highly excited.
The streams of lava grew to larger dimensions. It was
not, as is commonly imagined, a glowing, fluid mass, but
appeared like a stone wall, from twenty to thirty feet
high, consisting of vast blocks of stone, which were
partly black and partly glowing deep red, and this wall
was borne on high by the liquid, burning lava underneath, and pushed continually forward by the immense
weight of the fiery mass that issued forth from the cone.
Aided by the slope of the mountain side, it advanced
visibly some two or three feet a minute, threatening

death and destruction to all that stood in the way. A traveller, Mr. Boernstein, gives an animated description of a characteristic scene in its fearful progress. He had ascended the mountain as far as the *Casa del prete*, the priest's house, which was on the point of being over-whelmed by the stream of lava, now nearly four hundred feet wide.

It stood in the centre of a noble vineyard. The fur-niture, and all that could be saved, had been carried away; the old priest, in a roundabout and shorts, with nothing but his velvet skull-cap to designate him as a priest, was hard at work, with the help of a few men, to pull up the stakes to which the vines were fastened, in order to save them at least for fuel. His black dog was continually running towards the house, barking anxiously, and then returning to his master, barking at him and pulling at him, as if he wished to warn him against the impending danger. For the terrible wall of hidden fire was within a foot of the parsonage. It was empty and deserted; only a pet cat was sitting comfortably on the sill of the upper story, to which an outer staircase gave access. The priest had just cast a last sorrowful look at his house, against the thick stone wall of which the lava was slowly rising higher and higher, and in his heart was bidding farewell to his home, where he had lived ever since he had been a priest. His eye fell upon the cat. "Save the poor creature!" he cried, and one of the men hastened up the steps; but the cat, frightened by the strange face, ran swiftly into the house, and at the same moment the stream of lava, overtopping the house by several feet, fell over forward and poured a sea of flames upon the flat roof. The man on the steps leaped with a desperate effort to the ground; the priest and the by-standers crossed themselves; thick, black clouds of smoke poured forth from the windows; and a few˜minutes later the whole stately building had vanished, and a huge mass of glowing blocks of lava was steadily flowing over the place, that knew it no more.

At night the stream presents a glorious sight. Dark in broad daylight, it now appears an ocean of fire, slowly advancing with irresistible power; and from its waves, as high as tall houses, there fall continually huge glowing blocks with a fearful crash, and roll down the precipices with terrific thunder. If it approaches a tree, there is a moment's delay, and immediately the leaves, dried by the fearful heat, blaze up like a thousand lights on a huge Christmas tree; then the trunk flares up in a pillar of fire, and the crown sinks into the fiery sea. From time to time the glowing mass of the lava stream heaves and rises; suddenly a loud explosion is heard, and an immense column of bright fire shoots high up to the heavens—pent-up gases have freed themselves and exploded in the fiery heat. Or the stream falls into a well; the water is instantly converted into steam, and a white pillar of hissing vapor rises on high.

While Mount Vesuvius was thus relieving the Earth

in trouble, certain phenomena of smaller dimensions, but perhaps of greater interest even, were engaging the attention of the learned world. In the first days of the year 1866 the inhabitants of Santorin, an island in the Grecian Archipelago, had seen with amazement a part of their bay converted into a sea of fire. It was not that they had not witnessed the like before. Their own home is the result of a sudden upheaving of the bottom of the sea, and from time immemorial their bay has been the scene of fantastic transformations. The ancients spoke with awe of the strange changes that took place there— the island of Anaphi, now called Nanfi, rose at the bidding of Apollo from the lowest deep; Pliny mentions fearful convulsions, which marked the year 19 of our era, and ever since new islets have appeared and vanished again in the adjoining waters. Now, for nearly a year, subterranean thunders had been heard, and at the time mentioned tremendous explosions took place, red flames rose to the height of ten and twenty feet from the sea itself, and a few days later a new island ascended slowly, and grew visibly from hour to hour. The summit had the shape of a cone, and threw out an unceasing supply of stones, slime, and fire. During the following days, more islets presented themselves, and finally joined together, by means of the vast masses of half-fluid material that continued to flow from the crater. These new lands were nothing else but the summit of an immense volcano, which rested with its base on the bottom of the sea, while the summit, now for the first time, saw the light of day.

The power of man to accustom himself to any and every condition in which he finds himself placed, is strikingly illustrated by the ten or twelve thousand inhabitants of Santorin.   For three years now they have been living amid a continuous cannonade, surrounded by a sea on fire, and a volcano before their eyes, which does not cease day and night to throw out fiery projectiles amid heavy detonations.   Jets of vapor are sent up to a height of five thousand feet, and a perpetual fire illumines the top of Mount George I., as the new island has since been called.   Italian and Austrian engineers and savans from other countries, have been sent there to watch the extraordinary scene, and they report that the new island has risen already to a height of nearly five hundred feet, while it is still steadily increasing towards Santorin.   If the work continues at the same rate, the little kingdom of Greece has found out a cheaper means to increase its territory than the costly and dangerous process of annexation.   On the other hand, the apprehension has been expressed that in the bay of Santorin the waters of the sea may be deep enough to come in actual contact with the sea of fire in the interior of the earth, and that a fearful catastrophe may yet prove the old Greek doctrine of Hades and its horrors.

Whatever may be the true explanation of all these grave disturbances on our planet, whether we ascribe them, with the Neptunists, to the ebullition of heated waters, which seek an outlet, or with the Vulcanists, to the efforts of a sea of fire to break through the thin crust, and to hasten the day of final destruction, we cannot close our eyes to the fact that our mother Earth is evi-dently in trouble. But let us not blame her if blooming landscapes are laid waste, towns overthrown, and human lives sacrificed by hecatombs. The loss is great, the calamity appalling, but it is the price paid by a few for the security of the race. If the craters of volcanoes did not offer an opening to the pent-up vapors in the interior of the earth, and allow the terrific power of confined steam, with which we have of late become familiar in making steam our servant, we would not be able to live on the earth. They are, as already Alexander von Humboldt asserted, the safety valves, which allow the steam to escape, and the heated vapors within to regain their equilibrium with the pressure of the atmosphere; and it is thanks to· them only, perhaps, we owe, that we are enabled, by God's mercy, to enjoy our life on earth, although we dwell on a thin, frail crust, over an ocean of molten fire!

THE END.

A List of the Publications

OF

G. P. PUTNAM & SON,

661 *Broadway, New York*

ABBOT. MEXICO AND THE UNITED STATES.
Their Mutual Relations and Common Interests. With
Portraits on Steel of Juarez and Romero. By Gorham
D. Abbot, LL.D. 8vo, cloth, $3.50.

"This volume deserves the careful study of all true patriots. It has been prepared
with great care, from the most authentic sources ; and it probably furnishes the
materials for a better understanding of the real character and aims of modern
Romanism than any other volume extant."—*Congregationalist.*

AUDUBON. THE LIFE OF JOHN JAMES AU-
DUBON, the Naturalist. Edited by his widow. With
a fine portrait from the painting by Inman. 12mo,
cloth, extra gilt top, $2.50.

"It is a grand story of a grand life ; more instructive than a sermon, more romantic than a romance."—*Harper's Magazine.*

BASCOM. PRINCIPLES OF PSYCHOLOGY. By
John Bascom, Professor in Williams College. 12mo,
pp. 350, $1.75.

"All success to the students of physical science : but each of its fields may have
its triumphs, and the secrets of mind remain as unapproachable as hitherto. With
philosophy and not without it, under its own laws and not under the laws of a lower
realm, must be found those clues of success, those principles of investigation, which
can alone place this highest form of knowledge in its true position. The following
treatise is at least a patient effort to make a contribution to this, amid all failures,
chief department of thought."—*Extract from Preface.*

2 *Publications of*

BLACKWELL. STUDIES IN GENERAL SCIENCE. By Antoinette Brown Blackwell. 12mo (uniform with Child's " Benedicite "). Cloth extra, $2.25.

"The writer evinces admirable gifts both as a student and thinker. She brings a sincere and earnest mind to the investigation of truth."—*N. Y. Tribune.*

"The idea of the work is an excellent one, and it is ably developed."—*Boston Transcript.*

BLINDPITS—A Novel. [Reprinted by special arrangement with the Edinburgh publishers.] 1 vol. 12mo, $1.75.

** A delightful story, which everybody will like.

"The book indicates more than ordinary genius, and we recommend it unreservedly." —*Buffalo Courier.*

BOLTE (Amely). MADAME de STAEL ; A Historical Novel : translated from the German by Theo. Johnson. 16mo, cloth extra, $1.50.

[*Putnam's European Library.*]

" One of the best historical novels which has appeared for a long time."—*Illust. Zeitung.*

" Worthy of its great subject."—*Familien-Journal.*

" Every chapter brings the reader in contact with eminent personages, and entertains him in the most agreeable and profitable manner."—*Europa.*

" This is one of those valuable novels that combine historical and biographical information with amusement."—*Cincinnati Chronicle.*

BRACE. THE NEW WEST ; or, California in 1867 and '68. By Charles L. Brace, Author of the " Races of the Old World," " Home-Life in Germany," " Hungary in 1851," etc. 12mo, cloth, $1.75.

" We recommend it as the most readable and comprehensive book published on the general theme of California."—*N. Y. Times.*

BRYANT. LETTERS OF A TRAVELLER. By Wm. Cullen Bryant. New edition. 12mo, cloth.

—— LETTERS FROM THE EAST. Notes of a Visit to Egypt and Palestine. 12mo, cloth, $1.50.

—— The Same. ILLUSTRATED EDITION. With fine engravings on steel. 12mo, cloth extra, $3.

G. P. Putnam & Son. 3

AVÉ. THE CAVÉ METHOD OF LEARNING TO DRAW FROM MEMORY. By Madame E. Cavé. From 4th Parisian edition. 12mo, cloth, $1.

⁕ This is the *only method of drawing which really teaches anything.* In publishing the remarkable treatise, in which she unfolds, with surpassing interest, the results of her observations upon the teaching of drawing, and the ingenious methods she applies, Madame Cavé . . . . renders invaluable service to all who have marked out for themselves a career of Art."—*Extract from a long review in the Revue des Deux Mondes,* written by Delacroix.

" It is interesting and valuable."—D. HUNTINGTON, *Prest. Nat. Acad.*

" Should be used by every teacher of Drawing in America."—*City Item, Phila.*

" We wish that Madame Cavé had published this work half a century ago, that we might have been instructed in this enviable accomplishment."—*Harper's Mag.*

CAVÉ. THE CAVÉ METHOD OF TEACHING CO-LOR. 12mo, cloth, $1.

⁕ This work was referred, by the French Minister of Public Instruction, to a commission of ten eminent artists and officials, whose report, written by M. Delacroix, was unanimously adopted, endorsing and approving the work. The Minister, thereupon, by a decree, authorized the use of it in the French Normal schools.

G. P. PUTNAM & SON have also just received from Paris specimens of the MATERIALS used in this method, which they can supply to order. I. The GAUZES (framed) are now ready. Price $1 each. With discount to teachers. II. The Stand for the gauze. Price $1.50. III. MÉTHODE CAVÉ, *pour apprendre à dessiner* juste et de mémoire d'après les principes d'Albert Durer et de Leonardo da Vinci. Approved by the Minister of Public Instruction, and by Messrs. Delacroix, H. Vernet, etc. In 8 series, folio, paper covers. Price $2.25 each.

N.B.—The Crayons, Paper, and other articles mentioned in the Cavé Method may be obtained of any dealer in Artist's Materials. Samples of the French Articles may be seen at 661 Broadway.

HADBOURNE. NATURAL THEOLOGY; or, Nature and the Bible from the same Author. Lectures delivered before the Lowell Institute, Boston. By P. A. Chadbourne, A.M., M.D., President of University of Wisconsin. 12mo, cloth, $2. Student's edition, $1.75.

" This is a valuable contribution to current literature, and will be found adapted to the use of the class-room in college, and to the investigations of private students."—*Richmond Christian Adv.*

" The warm, fresh breath of pure and fervent religion pervades these eloquent pages."—*Am. Baptist.*

" Prof. Chadbourne's book is among the few metaphysical ones now published, which, once taken up, cannot be laid aside unread. It is written in a perspicuous, animated style, combining depth of thought and grace of diction, with a total absence of ambitious display."—*Washington National Republic.*

" In diction, method, and spirit, the volume is attractive and distinctive to a rare degree."—*Boston Traveller.*

HILD'S BENEDICITE ; or, Illustration of the Power, Wisdom, and Goodness of God, as manifested in His Works. By G. Chaplin Child, M.D. From the London edition of John Murray. With an Introductory Note by Henry G. Weston, D.D., of New York. 1 vol. 12mo. Elegantly printed on tinted paper, cloth extra, bevelled, $2 ; mor. ext., $4.50.

CHIEF CONTENTS.

Introduction.	Winter and Summer.	Wells.
The Heavens.	Nights and Days.	Seas and Floods.
The Sun and Moon.	Light and Darkness.	The Winds.
The Planets.	Lightning and Clouds.	Fire and Heat.
The Stars.	Showers and Dew.	Frost and Snow, etc.

"The most admirable popular treatise of natural theology. It is no extravagance to say that we have never read a more charming book, or one which we can recommend more confidently to our readers with the assurance that it will aid them, as none that we know of can do, to

'Look through Nature up to Nature's God.'

Every clergyman would do well particularly to study this book. For the rest, the handsome volume is delightful in appearance, and is one of the most creditable specimens of American book-making that has come from the Riverside Press."—*Round Table, N. Y.*, June 1.

LARKE. PORTIA, and other Tales of Shakespeare's Heroines. By Mrs. Cowden Clarke, author of the Concordance to Shakespeare. With engravings. 12mo, cloth extra, $2.50 ; gilt edges, $3.

⁎ An attractive book, especially for girls.

OOPER. RURAL HOURS. By a Lady. (Miss Susan Fenimore Cooper.) New Edition, with a new Introductory Chapter. 1 vol. 12mo, $2.50. ·

"One of the most interesting volumes of the day, displaying powers of mind of a high order."—Mrs. HALE's *Woman's Record.*

"An admirable portraiture of American out-door life, just as it is."—*Prof. Hart.*

"A very pleasant book—the result of the combined effort of good sense and good feeling, an observing mind, and a real, honest, unaffected appreciation of the countless minor beauties that Nature exhibits to her assiduous lovers."—*N. Y. Albion.*

RAVEN (Mme. Aug.). ANNE SEVERIN : A Story translated from the French. 16mo, $1.50.

[*Putnam's European Library.*]

⁎ "The Sister's Story," by the same author, has been warmly and generally eulogized as a book of remarkably pure and elevated character.

"By her great success, Mrs. Craven has larger power for good than perhaps any other writer in France."—*Pall Mall Gazette.*

ONTGOMERY. OUR ADMIRAL'S FLAG ABROAD; the Cruise of the U. S. Flagship Franklin, Admiral D. G. Farragut, in 1867-8. By James Eglinton Montgomery, of the Admiral's Staff. Illustrated with Drawings by Thos. Nast and Granville Perkins, from sketches by Lieut.-Commander Hoff and Park Benjamin, Jr., of the U. S. Ship Franklin, engraved in the best manner and printed in tints. In one volume, large octavo. Cloth, $7; morocco, $12. Also, cheaper edition, 8vo, cloth, $3.00.

ORTIMER. MARRYING BY LOT. A Tale of the Primitive Moravians. By Charlotte B. Mortimer. 1 vol. 12mo, $1.75.

In this volume will be found, in detail, ample illustrations of this extraordinary way of settling the matrimonial questions of the whole of a Christian Denomination.
" Good thoughts acted out in noble lives."—*Albany Argus.*
" A novel of excellent purposes and of considerable ability."—*Home Journal.*

'CONNOR. NETTIE RENTON; or, The Ghost: A Story. By W. D. O'Connor. With illustrations by Nast. $1.25.

"Rich in sweetness, pathos, and tender humanity."—*Providence Journal.*

ON THE EDGE OF THE STORM. By the Author of " Mademoiselle Mori." 12mo, cloth, with Frontispiece, $1.75. Also, the same book, cloth, gilt edges, adapted as a gift book for young ladies (ready Oct. 1st). $2.50.

"This is a charming story. The sympathy which the author evinces towards all her personages, and the justice she does to their different modes of thought and opinion, are the main charm of the book."—*London Athenæum.*
" This book is altogether a delightful one, showing great knowledge, a rare power of writing, and a far rarer artistic mastery over form and detail."—*Pall Mall Gazette.*

TIS. SACRED AND CONSTRUCTIVE ART. Its Origin and Progress. A series of Essays. By Calvin N. Otis, Architect. 16mo, cloth, $1.25.

"The work is eminently suggestive, and is written in a scholarly style, which commends it to the perusal of the cultivated student of literature and art."—*Boston Commonwealth.*

EPYS. Mr. Secretary Pepys and his Diary. Being a Sketch of the Times of Charles II. and James II. 12mo, cloth, $1.50.

Fay's Great Outline of Geography.

Tne Publishers take pleasure in announcing that FAY'S GREAT OUTLINE OF GEOGRAPHY is steadily gaining in popularity, and is being rapidly introduced in the best schools and colleges throughout the country. Members of Boards of Education, Superintendents of schools and colleges, and eminent teachers in public and private schools, as well as other distinguished friends of education, have everywhere presented themselves, impelled exclusively by the desire to examine a new and well-recommended school-book, and, if satisfied with it, to adopt it.

As a specimen of "press" and private Criticisms, we quote the following:

"For high schools and the best classes of grammar high schools, Fay's Great Outline is beyond anything that we have yet examined. Mr. Fay's atlas is incomparably the best thing in this country. In elegance and accuracy, the maps are well-nigh perfect. The best thing about Mr. Fay's manual is the plasticity which it permits; it can be used with much greater freedom from a slavish method than any other work that we have examined. It leaves a great deal to the ingenuity of the teacher; and in the hands of a living teacher it becomes a most elastic and expansive volume.

"We commmend the work heartily to the examination of our leading educators, and believe that it will stand the most trying of all tests, that of actual use in the school-room."—*The Courant, Hartford, March 27th,* 1868.

It has been approved by the Committee of the N. Y. Board of Education, and placed on the list of books to be used in the PUBLIC SCHOOLS OF THE CITY OF NEW YORK.

It is also used in many of the best Private Schools in MASSACHUSETTS, in NEW YORK, in OHIO, in ILLINOIS, and in KANSAS.

The press from all parts of the country has spoken earnestly, without other motive than interest in the cause of education. Telegrams and letters are continually received from States distant as well as near, encouraging publishers and author to persist in their efforts, and declaring that "after careful trial the book improves upon acquaintance—that the classes are interested and teaching is a pleasure."

Price of the Atlas and Text Book, in cloth, $3.75; half bound, $4. Liberal terms to Schools.

G. P. PUTNAM & SON, NEW YORK.

PUTNAM'S MAGAZINE.

NEW SERIES—THREE VOLS. COMPLETED.

"Calling into their service the best intellect of America, and making large original draughts upon the best known and most popular writers in Europe, the publishers have succeeded in constituting their monthly issues successive embodiments of the highest style of periodical literature, in every department. It is hardly possible to name an author whose works have rendered him familiar to the thinking public, who has not taken part in the task of rendering '*Putnam*' conspicuous among the best in an epoch which is producing so many good magazines. The three volumes of the new series contain two hundred and seventy-one original articles, and each of them is valuable in itself. These essays range the wide and prolific fields of biography, history, natural philosophy, scientific discovery and progress, art achievements, travel and exploration, religious opinion and controversy, poetry, sentiment, and romance.   *   *   *   *   *

"It was said of a famous scholar that to enjoy his acquaintance was a liberal education—and the same may be remarked of the habitual reading of '*Putnam*.' The admirable plan of this publication takes in all topics of modern thought and study; while every subject is invariably treated with ability, and so as to present clearly the best ideas it suggests. We are glad to know that the Magazine is a permanent and assured success as a business enterprise, and that the energetic publishers have now under contemplation a series of improvements that will still further advance their claims to the gratitude and patronage of those who can appreciate a high order of merit in literature."— * * *Albany Evening Journal.*

"In several features *Putnam* has been preëminent among the Monthlies. The first series was without a parallel in this country for the enduring value of its contents, and a score or more of standard volumes were made up from it. The new series has been equally admirable in this respect."—*Buffalo Com. Adv.*

"This really valuable Monthly continues to increase in each issue its contributions of lively interest and solid value."—*Charleston Courier.*

"It has taken its old place at the head of the Magazines of the day, and has kept pace with the constant growth of the literature of the period."—*Norwalk Gazette.*

The 3 vols. complete in cloth, comprising 2,200 pages of the choicest reading, will be sent free to any one sending us four subscribers with the money.

Three Months Free.—New subscribers for 1870, who send their names before 1st November, will be supplied with the last 3 numbers of the present volume FREE of charge.

Price $4.00 per annum; 2 copies for $7.00; 3 copies for $10.00. Liberal terms for Clubs, or with other periodicals.

G. P. PUTNAM & SON, PUBLISHERS,
661 BROADWAY, NEW YORK.

www.ingramcontent.com/pod-product-compliance
Lightning Source LLC
Chambersburg PA
CBHW021403210326
41599CB00011B/984